Southeast Asian Development

Southeast Asia has long fascinated development practitioners and researchers for being one of the few regions of the world that appears to be succeeding in its pursuit of development. Millions have emerged from poverty, international conflict is largely absent, and health and education indicators are improving. However this success can be questioned for concealing widespread hardships and for contributing to growing rates of inequality, suppression of civil and political rights, restructuring of local livelihoods, and widespread environmental degradation. The ability or otherwise of Southeast Asia, as one of the world's most successful developing regions, to overcome these challenges raises fundamental questions about the nature and desirability of development itself.

Divided into accessible thematic chapters, this book adopts the unique perspective of equitable development to outline the strengths and weaknesses of the transformations taking place in the Southeast Asian region. Four key themes are focused upon: equality and inequality; political freedom and opportunity; empowerment and participation; and environmental sustainability. These concepts are used to explore Southeast Asian development and trace the impacts that the growing popularity of market-led and grass-roots approaches are having upon economic, political and social processes. While the diversity of the region is emphasised so are some of the homogenising trends such as the concentration of wealth and services in urban areas and the subsequent migration of rural people into urban factories and squatter settlements. The ongoing commercialisation and industrialisation of rural agriculture, the expansion of non-farm income earning opportunities, and the alarming rates of environmental degradation that threaten health and livelihoods are also exposed.

In highlighting how Southeast Asian development is unevenly distributing wealth, opportunities and risks throughout the region, this book emphasises the need for creative new approaches to ensure that benefits of development are equitably enjoyed by all. Including illustrations, case studies and further reading, this book provides an accessible up-to-date introductory text for students and researchers interested in Southeast Asian development, development studies, Asian studies and geography.

Andrew McGregor is a senior lecturer in the Department of Geography at the University of Otago, New Zealand, and previously worked for UNICEF Australia. His research focuses on the culture, power and geography of foreign aid programmes in Southeast Asia.

Routledge Perspectives on Development

Series Editor: Professor Tony Binns, *University of Otago*

The *Perspectives on Development* series will provide an invaluable, up-to-date and refreshing approach to key development issues for academics and students working in the field of development, in disciplines such as anthropology, economics, geography, international relations, politics and sociology. The series will also be of particular interest to those working in interdisciplinary fields, such as area studies (African, Asian and Latin American studies), development studies, rural and urban studies, travel and tourism.

If you would like to submit a book proposal for the series, please contact Tony Binns on j.a.binns@geography.otago.ac.nz

Published:

David W. Drakakis-Smith
Third World Cities, Second Edition

Kenneth Lynch
Rural–Urban Interactions in the Developing World

Nicola Ansell
Children, Youth and Development

Katie Willis
Theories and Practices of Development

Jennifer A. Elliott
An Introduction to Sustainable Development, Third Edition

Chris Barrow
Environmental Management and Development

Janet Henshall Momsen
Gender and Development

Richard Sharpley and David J. Telfer
Tourism and Development

Andrew McGregor
Southeast Asian Development

Forthcoming:

Jo Beall
Cities and Development

Hazel Barrett
Health and Development

Cheryl McEwan
Postcolonialism and Development

Tony Binns, Christo Fabricius and Etienne Nel
Local Knowledge, Environment and Development

Andrea Cornwall
Participation and Development

Janet Henshall Momsen
Gender and Development, Second Edition

Tony Binns and Alan Dixon
Africa: Diversity and Development

Andrew Williams and Roger MacGinty
Conflict and Development

Michael Tribe, Frederick Nixon and Andrew Sumner
Economics and Development Studies

David Lewis and Nazneen Kanji
Non-Governmental Organisations and Development

Clive Agnew and Philip Woodhouse
Water Resources and Development

David Hudson
Global Finance and Development

W.T.S. Gould
Population and Development

Andrew Collins
Disaster and Development

Southeast Asian Development

Andrew McGregor

LONDON AND NEW YORK

First published 2008
by Routledge
2 Park Square, Milton Park, Abingdon, Oxon OX14 4RN

Simultaneously published in the US and Canada
by Routledge
270 Madison Ave, New York, NY 10016

Routledge is an imprint of the Taylor & Francis Group, an informa business

Typeset in Times New Roman by
RefineCatch Limited, Bungay, Suffolk
Printed and bound in Great Britain by
Antony Rowe Ltd., Chippenham, Wiltshire.

British Library Cataloguing in Publication Data
A catalogue record for this book is available from the British Library

Library of Congress Cataloging in Publication Data
McGregor, Andrew, 1971–
Southeast Asian development / by Andrew McGregor.
p. cm.
Includes bibliographical references and index.
1. Southeast Asia–Economic policy. 2. Southeast Asia–Economic conditions.
3. Southeast Asia–Economic social conditions. I. Title.
HC441.M42 2008
338.959–dc22
2007040840

ISBN 10: 0–415–38416–8 (hbk)
ISBN 10: 0–415–38152–5 (pbk)
ISBN 10: 0–203–08600–7 (ebk)

ISBN 13: 978–0–415–38416–2 (hbk)
ISBN 13: 978–0–415–38152–9 (pbk)
ISBN 13: 978–0–203–08600–1 (ebk)

For my whanau,

Jack, Charlie, Rachel, Christian, Diana, Mum and Dad

Contents

Figures

Tables

Boxes

Acknowledgements

This book could never have been written without the inspiration, guidance and support of a range of wonderful people. First of all I would like to thank series editor Tony Binns for approaching me and expressing confidence in my abilities despite my own personal reservations and doubts. His enthusiasm and belief helped me complete the book as have the patience and encouragement of Andrew Mould and Jennifer Page from Routledge. For different reasons I would like to thank Simon Stroud, Gaye Phillips, Sarah Lendon and the rest of my colleagues at UNICEF Australia who first introduced me to the challenging world of foreign aid and development, and remain inspirations through their tireless commitment. Academically I would like to thank my colleagues in the Department of Geography and the Poverty, Inequality and Development cluster at the University of Otago for providing a stimulating work environment as well as my old University of Sydney colleagues, particularly John Connell, Philip Hirsch and Chris Gibson, for fulfilling a similar role there. My students deserve praise for their consistent enthusiasm and stimulating ideas with special thanks going out to Sarah Ellis, Jessica Hattersley, Gradon Diprose, Luke Swainson, Sunita Devasahayam, Tim Lester, Luke Roper, Hamish Saunders and Stephen Molloy. Much of this book was written while on sabbatical for which I thank the University of Otago as well as the National University of Singapore, particularly Shirlena Huang, Tim Bunnell, Assi Doran and Lai Wa, for hosting me and ensuring my trip was so productive and enjoyable, as well as Dani and his family for hosting me with such warmth in Aceh.

Deserving special praise for contributing boxes are Rebecca Miller from the University of Auckland, Brody Lee from the University of Otago and my former students Luke and Gradon. Thanks also go out to Dennis Viera, Sarah Ellis, Jane Dunlop and Kara Barnett for contributing photos and to Tracy Connolly for putting together some excellent illustrative materials. The three reviewers deserve thanks for providing some very useful critical feedback that improved this final version. The most important 'silent' partner throughout the writing process has been Guil Figgins who has played a vital role in drafting many of the boxes and figures, helping out with the references, and being a constant source of friendship and encouragement. In a broader sense I thank all those who have been involved in my research over the years for inspiring many of the ideas written in the pages that follow. To my whanau and friends in Dunedin and Sydney, thanks for all the encouragement and good times during the writing of this book, and to my old friend Alex Lesuik for staying forever young. Finally, I would like to thank Kara Barnett for her belief, encouragement, love and unwavering enthusiasm through some busy and stressful times; you are my inspiration.

Abbreviations

AFTA	ASEAN Free Trade Agreement
ASEAN	Association of Southeast Asian Nations
CARP	Comprehensive Agrarian Reform Programme
CMP	Community Mortgage Programme
CPP	Cambodian People's Party
EOI	export-oriented industrialisation
EMR	extended mega-urban region
EPZ	export processing zone
EU	European Union
FAO	Food and Agriculture Organisation
FDI	foreign direct investment
FELDA	Malaysia Federal Land Development Authority
GDP	gross domestic product
GM	genetically modified
GNP	gross national product
GONGO	government organised non-government organisation
HDI	Human Development Index
HYV	high yielding variety
IMF	International Monetary Fund
IMS-GT	Indonesia–Malaysia–Singapore Growth Triangle
IPMP	Integrated Pest Management Programme
IRRI	International Rice Research Institute
ISI	import substitution industrialism
JI	Jemaah Islamiah
MDG	Millennium Development Goal

MRC	Mekong Rivers Commission
NEM	New Economic Mechanism
NEP	New Economic Policy
NGO	non-governmental organisation
NLD	National League for Democracy
PAP	People's Action Party
PAS	Pan-Malaysia Islamic Party
RECOFT	Regional Community Forestry Training Centre for Asia and the Pacific
SAP	structural adjustment package
SARS	Severe acute respiratory syndrome
SEAFDC	Southeast Asian Fisheries Development Centre
SLORC	State Law and Order Restoration Committee
SSP3	Sungai Selangor Water Supply Scheme Phase III
TNC	transnational corporation
UNDP	United Nations Development Programme
UNICEF	United Nations Children's Fund
UNMO	United Malays National Organisation
UNTAET	United Nations Transitional Administration in East Timor
VOC	United East Indian Company

1 Introducing Southeast Asian development

Introduction

Southeast Asia is often portrayed as a development success story. With the development industry attracting increasing disillusionment and critique for its modest achievements in overcoming global economic inequality Southeast Asia shines like a beacon of hope that suggests development can work and that conditions can improve. There can be little doubt that Southeast Asia has achieved some incredible things in a relatively short period of time. Living standards have improved; economies, despite the Asian economic crisis, are generally strong and growing; international peace has descended upon the region; and millions of people have emerged from poverty. Southeast Asia compares favourably to almost any other developing region in the world and the prospects for the future are good. However the benefits of development have not been distributed evenly throughout the region. Instead some countries are extremely wealthy while others struggle to feed their populations; urban centres inevitably outperform rural areas in access to wealth and services; and even within communities development often favours the more educated and able over the extremely poor. Southeast Asia is successful but this success is not enjoyed by all, instead the benefits experienced by some are causing hardship and difficulty to others. This book explores these dilemmas that lie at the heart of Southeast Asian development by focusing upon the themes of equality and inequality, political freedom and opportunity, community participation and empowerment, and environmental sustainability. These themes reveal the challenges and opportunities facing more 'successful' and more equitable forms of Southeast Asian development.

In this book development is conceived to be the array of programmes, approaches and policies that are introduced by a variety of local, state and international actors with the intention of bringing about positive or good change in the quality of people's lives. Traditionally such change has focused on the economy, looking at ways of boosting national and individual incomes with the belief that additional wealth equates to improvements in quality of life. As will be described in more detail below, such beliefs have been disputed and development initiatives now target a wide range of sectors, some focusing on social change through health and education, others upon political change through good governance schemes, and others concentrating upon environmental sustainability through conservation programmes. While all aim to benefit society and most do, they are inherently controversial because of the uneven distribution of costs and benefits and the uneven access people have to decision-making processes. While development actors are continually reflecting upon and attempting to improve their practices, the sheer size and ambitious nature of their endeavour – to bring about positive change – ensures development will maintain its contested nature for years to come.

One of the core struggles surrounding development derives from the shifting expectations and definitions different states, social groups and communities have of positive change and therefore what constitutes good and effective development. This is particularly the case in Southeast Asia, which hosts one of the most diverse populations of any region in the world, multiplying the number of ways in which development might be approached and appreciated. There are eleven states in Southeast Asia, home to some 600 million people (see Figure 1.1). Each state boasts multiple ethnic groups, languages and cultures with population sizes varying from 370,000 people in the tiny oil-rich sultanate of Brunei Darussalam, to 217 million people sprawled across the Indonesia archipelago, the fourth most populous country in the world (see Table 1.1). The world's biggest religions, Islam, Buddhism, Christianity and Hinduism, are all represented in the region as are Confucian teachings and community-based animist belief systems not practised anywhere else in the world. Economically, the region could hardly be more diverse with Singapore and Brunei ranked among the world's twenty richest countries while citizens of five other countries average less than US$2 a day. Such discrepancies are heightened within

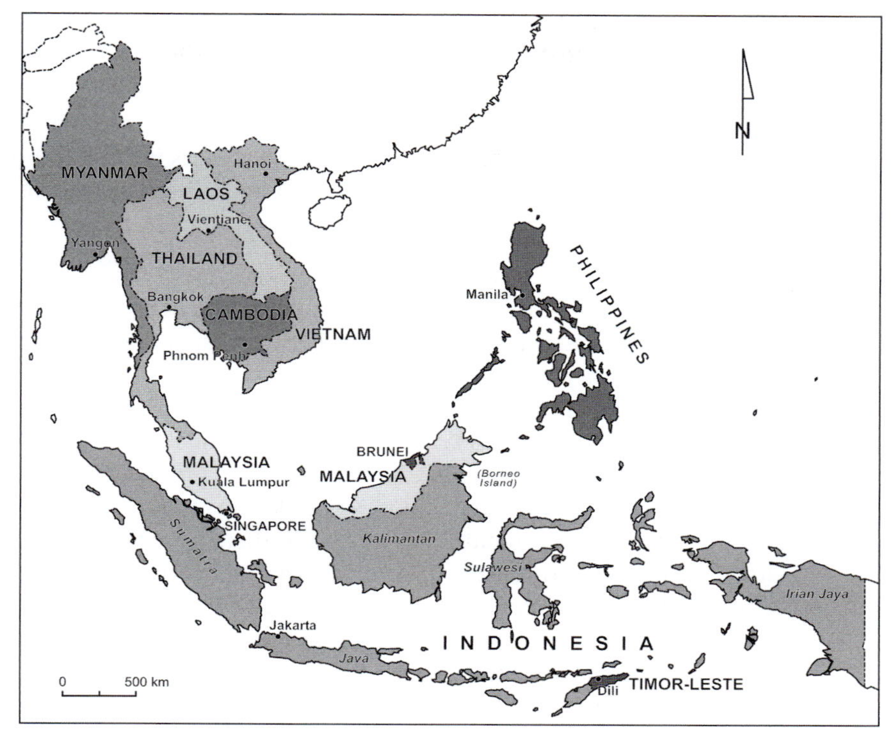

Figure 1.1 Southeast Asia.

Table 1.1 ***Diversity within Southeast Asia***

	Population 2006 (millions)	*Life expectancy 2004 (years)*	*Infant mortality 2004 (per 1,000)*	*GDP per capita 2005 ($US)*	*HDI 2004 (world rank)*
Singapore	4.46	78.6	3	26,821	25
Brunei Darussalam	0.38	76.3	8	16,882	34
Malaysia	26.68	73	10	5,001	61
Thailand	65.23	69.7	18	2,726	74
The Philippines	85.85	70.2	26	1,160	84
Indonesia	222.73	66.5	30	1,275	108
Vietnam	84.22	70.4	17	635	109
Cambodia	14.0	56	97	404	129
Myanmar	57.29	60.1	76	106	130
Lao PDR	6.14	54.5	65	418	133
Timor Leste	0.89	55.2	64	367	142

Sources: UNDP (2006), ASEAN (2006), UNESCAP (2006)

borders with urban elites often having access to the latest and most expensive consumer goods while some of their more remote rural cousins are yet to access electricity. Political systems range from open liberal democracies to repressive military-ruled states with some countries hosting a free and active media while others imprison those who voice alternative views. Even the natural landscapes are incredibly varied from the snow-tipped mountains of Myanmar[1] to the rich tropical rainforests on the island of Borneo to the long sandy beaches and mangrove swamps of Indonesia.

Such diversity is unusual within a single region, which makes it seem likely that proximity, rather than any shared characteristics or identity, drove the rationale for labelling the disparate collection of states together as a region. Rajah (1999), for example, suggests the concept of Southeast Asia has resulted more from interaction and recognition of shared interests among the region's populations rather than any essentialist similarities in nature or society. Others have traced the history of the term Southeast Asia to the Second World War when the British used it as convenient shorthand to refer to military operations occurring in the area (Emmerson 1984). However there can be little doubt that Southeast Asian states have since acted to cement the idea of a regional identity through what Acharya (1999) refers to as a 'collective act of self-imagination' to access the economic, political and geo-strategic advantages that come with it. The evolution of the Association of Southeast Asian Nations (ASEAN), which now counts all countries apart from the recently independent Timor-Leste (which has observer status) as members, is the most apparent and outwardly obvious manifestation of this emerging regional identity. However these new structures of political unity and identity have contributed little to a unified vision of development that differs between and across states as well as within communities and even households. To come to a common understanding of good and positive change among such diversity is quite impossible.

The divergence in development visions is readily apparent at the state scale. Singapore, for example, has focused upon transport infrastructure and export-oriented industrialisation to successfully

[1] Burma was renamed Myanmar by the ruling military government in 1989. As a form of protest against the government many Western countries and Burmese activists refuse to recognise this new name. However the name is accepted elsewhere, including within the United Nations and ASEAN. This book refers to the territory as Burma when referring to specific incidents from independence to 1989, and Myanmar for more recent events and as a general reference.

attract large amounts of foreign direct investment (FDI) that have afforded its residents with lifestyles unparalleled, in terms of services and income, anywhere else in the region. In contrast Myanmar has pursued nationalistic visions of development through policies of isolationism, a project that has had largely negative ramifications for the quality of life of its citizens. Vietnam and Laos have put their faith in socialist state-controlled economies while the Philippines, Thailand, Indonesia and Malaysia have all been much more market-oriented. Brunei has relied almost entirely upon oil exports to fund its development strategies, a commodity that is likely to be of immense importance to the future of the newly independent Timor-Leste. Finally, Cambodia has suffered the most due to internal rifts about appropriate development directions, experiencing years of civil war and millions of deaths as various factions fought for power and attempted to implement their visions of positive change and appropriate development.

Conflicting views on development are also evident at smaller scales, in provinces, communities and households. The simple process of building a road, for example, can be steeped in controversy at the local level. For some a road is a symbol of development and provides a means of transporting the benefits of development such as medicines, foods and new technologies to remote communities while also providing infrastructure for distant peoples to migrate, trade and otherwise interact with one another. For others, however, a road can represent unwanted intrusion and change by facilitating the import of externally produced goods that destroy local markets and livelihoods or by hastening the migration of young people away from their traditional homelands. Roads may also provide means for states to enhance their surveillance and control of marginalised communities. In other words good change from one perspective can equate to negative change from another. In Southeast Asia development has seen a reduction in the number of people living in poverty but an increase in economic inequality; an improvement in the quality of houses but a spread of squatter settlements and slums; an increase in rural land values but an increase in landlessness; and the proliferation of protected national parks but an increase in deforestation. Development affects people and places in different ways and it is these contrasting impacts and paths of development within Southeast Asia that are the focus of this book. While the book is cautiously supportive of development, it is also critical, and seeks to generate awareness and appreciation for important concepts such as equality, opportunity, participation and sustainability. To understand the importance of these concepts key development theories and perspectives will be introduced in the context of Southeast Asia.

Perspectives on development

There is no widely accepted definition of development. Instead, as is the case in this book, most academics and practitioners outline their own understanding of the concept and use this as a lens to interact with and interpret the subjects they are examining. The concept of development is often loosely used to refer to progress or evolution, describing the emergence of ever more sophisticated civilisations, particularly since the European Enlightenment. Development in this generalist sense is seen as moving forward or advancing from one form to another. In contrast Cowen and Shenton (1996) suggest that development has evolved in opposition to progress, to mitigate its negative effects. Progress implies change and change destroys that which existed before it, sometimes with painful impacts, such as unemployment or marginalisation, for particular groups within society. Development, in this view, positions developers as trustees who implement programmes and projects in the interests of people disadvantaged and disenfranchised by progress. The first generalist perspective sees development as broad processes of change or evolution, whereas the second trusteeship perspective sees development as a specific set of projects embedded with power relations and reacting to the destructive impacts of change. These concepts evolved in Europe but played an important role in legitimising early colonial incursions into the Southeast Asian region. Colonialists could use the generalist definition to argue they were assisting in the evolution or progression of Southeast Asian societies, and the trusteeship perspective to implement projects and initiatives that helped them believe they were acting in the interests of their Southeast Asian subjects. Hence while Europeans were extracting land, labour and resources from the region they were able to morally legitimise their actions by promoting the concept of development.

Development has become much more institutionalised and visible since the end of the Second World War. It was then that colonial empires began to collapse and a new system for modelling international relations between independent states was required. To this end an important conference took place at Bretton Woods in the US that resulted in the establishment of the World Bank, the International Monetary Fund (IMF) and the General Agreement on Tariffs and Trade (later to be the World Trade Organisation). These three institutions assisted in running and stabilising the world economy while the United Nations, which formed in 1945, was set up to pursue peace and protect human rights. At the same time the US initiated its Marshall Plan that transferred funds and expertise from the US to help rebuild Europe. This process has since formed the blueprint for subsequent interventions, in the form of foreign aid, from Western nations into regions such as Southeast Asia. The formalisation of these institutions and interventions signalled the starting point for the current phase of development and the establishment of identifiable development institutions and industries. Many point to President Truman's inaugural speech in 1949 as the moment from which development grew to take its contemporary form. He spoke out against 'the old imperialism – exploitation for foreign profit' and instead outlined a 'program of development' to make 'the benefits of our scientific advances and industrial progress available for the improvement and growth of underdeveloped areas' (Truman 1949).

This flurry of activity at the end of the Second World War heralded the beginning of a new phase of international relations in which development was articulated as the process and goal of foreign and domestic policies. Throughout Southeast Asia leaders professed their desire to develop their countries while international donors, particularly the US and the USSR during the Cold War, legitimised their programmes in development terms. New knowledges, languages, expertise and modes of interaction arose that portray development as both universally desirable and achievable. Some consider development to now be a meta-discourse that has bound the world into a collective struggle 'to develop', infiltrating and shaping everything from global economic interactions and national policy and resource allocations, to household relationships and individual imaginaries. Indeed it is now difficult to imagine a world without development. Yet despite the concept's popularity its definition remains elusive, slipping between different meanings for different people and places and changing over time. These diverse meanings can be seen in the evolution of development theories that have progressed from their early economic roots to a much broader range of concerns and interests.

Economic development theories

In the immediate post-war period development became closely associated with economic growth. A country with a stronger economy was seen as better or 'more developed' than one with a weak economy and 'to develop' referred to strategies of enhancing the economic output of a place. At the international scale the key unit to compare and rank countries in terms of their development was gross domestic product (GDP), which refers to the total value of the goods and services produced by that country each year. GDP is normally divided by population for a GDP/per capita ratio allowing the economic efficiency of each country per person to be calculated. The invention of this type of measurement system necessitated the proliferation of particular economic knowledges and expertise that allowed for GDP to be identified and calculated, which then formed the knowledge base of state development strategies. The most popular early development strategies were derived from the collection of ideas that came to be known as modernisation theory. Championed most famously by Walt Rostow, a US-based academic and policy-maker, his 1960 book, *The Stages of Economic Growth: A Non-Communist Manifesto* (Rostow 1960), became a key conceptual framework for modelling development in the early post-war period. According to Rostow countries develop along a linear path of five distinct stages from 'traditional society' to the final 'age of mass consumption'. Traditional societies were stereotyped as primitive and backward, defined by their absences or what they lacked when compared to the final stage which was modelled on the structure of US society. Development under modernisation theory became oriented towards replicating the economic and political structures of the US in other parts of the world.

Modernisation theory has since fallen from favour in development circles for being overly descriptive, rather than prescriptive, and for an overtly US bias. In grouping large swathes of the world, including Southeast Asia, under the heading of 'traditional society' there was no appreciation or respect for the diversity of the world's civilisations, all of which were positioned negatively and in need of modernisation. In Southeast Asia, for example, the ancient and sophisticated Confucianist Vietnamese state was lumped together with small animist villages scattered across the Philippines as 'traditional society'. The suggestion that all countries should undergo modernisation showed little awareness about the particularities of place and wrongly positioned the US as a more desirable and advanced endpoint than other places. Despite its faults modernisation theory was very influential with US policy-makers who found political mileage in Rostow's claim that communism would prevent countries progressing from one stage to the next, so in effect, communism was anti-development. Modernisation ideas also came to influence the Bretton Woods institutions and the United Nations which would use their power to encourage others to pursue modernisation goals. In Southeast Asia early US and multilateral support for the original ASEAN-5 nations (Singapore, Malaysia, Thailand, the Philippines and Indonesia see Chapter 3) was based on modernisation principles while the US invasion of Vietnam was partly premised on the belief that communism was unfairly preventing people there from developing (see Menzel 2006). While modernisation has since fallen from favour many of the ideas naturalised during this period continue to linger and influence development decision-making today.

Modernisation purported that the root of a country's underdevelopment lay within the country itself – once these faults were rectified the country could make the transition into a more developed state. This didn't ring true for many people who, having witnessed the destructive tendencies of colonialism, feared that the new international systems evolving under modernisation might have equally destructive impacts for poorer states. Concerns that development might evolve into a form of neo-colonialism saw neo-Marxist theories, such as dependency theory, arise in the 1960s and 1970s. Neo-Marxist economists such as Raul Prebische and Andre Gunder Frank believed that global trade relations were the greatest hindrance to development rather than the internal make up of states themselves. These theories positioned the world in two camps: the core, which was predominantly North America and Europe; and the periphery, which was the rest of the world. Peripheral countries were seen as locked into exploitative relationships with the core as their economies were based on selling cheap raw materials to the core which would add value to these products by processing them before consuming them or selling them back to the periphery at much higher prices. While early classical economic theories suggested that such an arrangement was mutually beneficial, neo-Marxist writers argued the system was self-perpetuating and was producing an international division of labour that favoured core countries. Those in peripheral countries were locked into low income employment in primary industries, with few opportunities to progress, while core countries were developing skilled secondary and tertiary employment sectors that produced high value goods and services stimulating further economic development and innovation.

A key insight from neo-Marxism is that the relative value of exporting manufactured products from the core to the periphery increases over time whereas the value of exporting primary products from the periphery to the core diminishes, creating an ever-widening gap between the wealthy economies of the core and the less wealthy periphery (for an introductory summary see Kay 2006). The solution to this problem, from a dependency theory perspective, is to industrialise peripheral economies by introducing a series of tariffs, taxes and subsidies that create favourable conditions for domestic investment and reduce dependence on the core. Through neo-Marxist theories development became less about a country progressing through evolutionary stages and more about the ability of a society to industrialise and generate secondary and tertiary industries – often by minimising their reliance upon the global economy. As will be discussed in Chapter 3 dependency theories were very influential in Southeast Asia with a variety of industrialisation strategies being adopted. These successfully grew local industries that would produce manufactured goods for domestic markets, however these markets were quickly saturated challenging the long-term logic of such approaches. The biggest challenge for neo-Marxist theory, however, came in explaining the success of Singapore, and the other Asian tiger economies of Taiwan, Hong Kong and South Korea. With small domestic markets neo-Marxism has struggled to explain their success as they have clearly benefited from international trade. World systems theory has suggested such countries should be categorised as the semi-periphery, playing an important bridging role between core and periphery. Semi-peripheral countries are seen as sharing many of the same interests as core countries, thereby contributing to the perpetuation of an unbalanced world system.

Neo-Marxist theories emerged from Latin America and, as they place blame for underdevelopment on the exploitative systems of core countries, they have always been more popular in the periphery than the core. In the 1980s the core reacted with what is now the most common international economic development theory, neo-liberalism. Popularised by British Prime Minister Margaret Thatcher and US President Ronald Reagan neo-liberalism argues that the cause of underdevelopment is the lack of freedom accorded to financial markets. Under a neo-liberal model of development government intervention into market activities, in the form of tariffs, subsidies or other taxes and protections, would be removed as they are perpetuating uncompetitive industries. In the absence of government intervention, it is argued, free markets will flourish and capital will naturally flow to areas of competitive advantage. For example, someone wishing to build a labour-intensive garment-production factory will build in a country where labour costs are low, such as Vietnam, if they are free to do so. Over time these investments will boost the economy of the low income country so that it will begin to 'catch-up' to wealthier economies. Similarly high wage countries would have to reduce their wages to compete with low income countries if they want to attract the same investment. In this way unimpeded market forces are expected to redistribute resources fairly, reducing the role of government in neo-liberal societies to the facilitation of market forces.

Unlike neo-Marxist theory neo-liberal approaches put the blame for underdevelopment squarely upon the policies of governments in poorer countries. The World Bank, IMF and World Trade Organisation along with core aid donors are all involved in pressuring less developed countries to minimise government intervention within the economy. If a country is in crisis and needs to borrow money from the IMF they are forced to adopt neo-liberal policies as a condition of borrowing. This was the case for Indonesia during the Asian economic crisis when it was forced to undergo a 'short sharp shock' and apply structural adjustment policies as a condition of borrowing much needed money from the IMF. In removing protections from uncompetitive industries and shrinking the number of government personnel by privatising government services, unemployment numbers soared as the prices of basic goods increased, propelling millions of people into poverty. There are many critics of neo-liberalism who argue, among other things, that the absence of government intervention encourages a 'race to the bottom' as wages and workers' rights are depressed in order to attract international capital; that privatisation of government services leads to increasing costs for consumers rather than more competitive pricing; and that allowing the free movement of international capital enhances the opportunities of the core to exploit the periphery causing social and environmental degradation. In response neo-liberalists have pointed to countries such as Singapore and Malaysia as examples of successful neo-liberal development, a severe misrepresentation, given the historically strong role governments have played in protecting domestic economies. Despite this most Southeast Asian governments are now involved in liberalising their economies in pursuit of economic development (see Chapter 3).

Broadening development

In the latter part of the twentieth century there have been moves away from simply equating development with economic growth and GDP to incorporate a much wider array of non-economic variables. The broader focus has derived from the realisation that boosting a country's GDP will not necessarily bring an increase in the quality of life for all. Economic gains, for example, may only benefit a wealthy urban merchant class or corrupt political elite while bringing few improvements, or even detracting from, the quality of life of the majority. This can be seen in the develop of large dams in Southeast Asia that boost a country's overall GDP but cause flooding and irreversible hardship for those forced to leave their homes and livelihoods. In recognition of these issues the concept of development has been broadened to include social, cultural, political and environmental concerns and conditions. Development is now practised in different ways, with a wide range of non-economic knowledges derived from subjects such as geography, anthropology, sociology and political science, becoming important.

One of the clearest examples of this shift in thinking at the international scale was the United Nations Children's Fund publication entitled *Adjustment with a Human Face* (Cornia *et al.* 1987). Within this document UNICEF critiqued the free market models promoted by the IMF for their negative impacts upon society, effectively questioning whether development should be primarily about raising GDP or whether social wellbeing is more important. This type of thinking became mainstreamed when the United Nations Development Programme (UNDP) began publishing its annual Human Development Reports in 1990 (UNDP 1990). Within these reports every country in the world is rated according to a diverse series of indices, both economic and non-economic, to identify their ranking within a global Human Development Index (HDI). While GDP is included so are social indicators such as life expectancy, literacy, school enrolments and gender equity. These new categories, which were traditionally marginal to most mainstream development debates, have been repositioned as central. This shift can be seen in the Millennium Development Goals (MDGs) which were signed off in 2000 and comprise a unified statement by the international community about the direction of development (see Table 1.2). The MDGs identify eight key goals for signatory states to work towards by 2015, including things such as improving maternal health, but making no mention of raising GDP. Development is being reconstructed as enlargening people's

Table 1.2 ***Millennium development goals and targets***

Goals	Targets – from 1990–2015
1 Eradicate extreme hunger and poverty	• Halve the proportion of people living on US$1 a day • Halve the proportion of people suffering from hunger
2 Achieve universal primary education	• Ensure all children are able to complete primary school
3 Promote gender equity and empower women	• Eliminate gender disparity in primary and secondary education
4 Reduce child mortality	• Reduce by two-thirds the under five mortality rate
5 Improve maternal health	• Reduce by three-quarters the maternal mortality ratio
6 Combat HIV/AIDS, malaria and other diseases	• Have halted and begun to reverse the spread of HIV/AIDS, malaria and other diseases
7 Ensure environmental sustainability	• Integrate principles of sustainable development into country policies and reverse loss of environmental resources • Halve the proportion of people living without access to safe water and sanitation • Achieve a significant improvement in the lives of at least 100 million slum dwellers (by 2020)
8 Develop a global partnership for development	• Develop an open, rule-based, predictable, non-discriminatory trading and financial system • Address the special needs of least developed countries, landlocked developing countries and small island developing states • Address debt problems to make debt sustainable in the long term • Develop strategies for decent and productive work for youth • Provide access to affordable essential drugs • Share benefits of new technologies, particularly information and communications

choices through boosting life expectancy and health, increasing knowledge and awareness through education, encouraging economic and political freedoms, and heightening access to resources that enable a 'decent standard of living'. While income and economic growth are important components of this process they are not the sole components, repositioning a wide range of non-economic variables as central to development.

An outcome of this broader view of development has been a greater awareness of diversity and difference. The 'one size fits all' approach of economic theorists has been replaced by an appreciation of geography – different places develop in different ways. This awareness is reflected in alternative or grassroots development theories that promote bottom-up decision-making processes steered by the people being 'developed' rather than external development experts. This is in contrast to most mainstream economic development approaches that explicitly or implicitly encourage top-down decision-making models through their reliance on macro-economic data. Grassroots development practitioners are much more likely to visit villagers to discuss their priorities and problems in person. Underlying the grassroots approach is a belief that the key to improving local livelihoods lies with the villagers themselves and the role of the development apparatus is to facilitate the realisation of these capacities. Consequently grassroots development focuses on the local scale with research focusing on improving the point of contact between the development industry and local communities. New methodologies have been explored that that encourage participation, empowerment, indigenous knowledge and local ownership.

Grassroots approaches have been criticised for being too focused on particular localities and for diverting attention away from broader economic, social and political processes. In addition the participatory approach is time- and labour-intensive, which slows development processes and makes them less attractive to donors looking for quick results. Host governments may also be resistant to such approaches as they are often cut out of the loop, reducing the opportunity for them to benefit from development exchanges while also eroding their authority as villagers become more self-reliant or, in some cases, more aid-reliant. Nevertheless grassroots approaches are becoming more and more popular within Southeast Asia, particularly among non-governmental organisations (NGOs) that are opposed to the more destructive elements of neo-liberalism. Indeed it is now rare for any development institution, whether it be the World Bank or a small Marxist NGO, to not at least claim their projects are participatory and driven by the grassroots.

More recently a controversial collection of ideas has arisen that is extremely critical of development and the development industry. Known as post-development theorists these writers draw upon post-structural theory to argue that development has caused more harm than good. Post-development writers argue that development is a discourse being used by the developed world to rearrange the developing world in their interests. Rather than liberating poorer countries, development is portrayed as a myth that has dislocated people from their cultures, lands, spirituality and traditions as they unsuccessfully pursue the consumptive lifestyles of rich countries. Only a small selection of elites actually achieve this end while the vast majority become stuck within a global production system that severs them from their past while promising them an unachievable future. While Southeast Asia is rarely cited by post-development writers, as many of the arguments appear to have more resonance in regions of Africa or South America, they are still very relevant critiques that can be applied to the region. The mass movement of people from rural areas to work in urban factories, for example, could be portrayed as people swapping their spiritually rich rural cultures for poorly paid, monotonous and often hazardous work in 'modern' industrial practices (problems with this portrayal are discussed in Chapter 6). Similarly popular Filipino resistance to the introduction of genetically manipulated crops is derived partly from fears about the impacts this will have upon rural traditions of seed sharing and collective harvesting – processes that are integral to community identities (see Chapter 7). For post-development theorists development is a myth that has unnecessarily ruined the lives of people whose imaginations have been seduced by its promise and appeal.

While post-development is strong on critique it is light on direction. Some writers have grandly called for an abandonment of the development project yet give few indications of how this could happen. Others, such as Escobar (1995a, 1995b), make more useful contributions by calling for *alternatives-to-development*, ways of moving into the future that are not reliant on development professionals and industries and not based on pre-determined norms, rankings and categories such as the MDGs. Implementing or encouraging alternatives-to-development is problematic, particularly given that the pursuit of mainstream development is so entrenched throughout the world. Post-development theorists put their faith in civil society institutions such as social movements, believing they provide the necessary socio-political space where people can express and explore indigenous alternatives-to-development. As such contact with traditional development institutions, such as the Bretton Woods institutions, are minimised although there is still a role for post-development-oriented NGOs to ferment, support and encourage alternative imaginaries. While such an approach may seem good in theory it faces many hurdles in practice, not least the fact that most social movements form to contest the distribution of developmental benefits rather than question the appropriateness of development itself. In Southeast Asia civil society is often repressed, limiting the potential of this sector to contribute alternatives-to-development (see Chapter 5). Nevertheless experiments and debates that are critical of development are taking place at the local scale across the region both in and out of view of the state and the broader development apparatus (see Gibson-Graham 2005; McGregor 2007). The first step, from a post-development perspective, is to break down the language and metaphors of development that confine our thinking to allow creative alternatives to be imagined.

Perceiving Southeast Asian development

Development clearly means different things to different people and many books could and have been written on the theme 'Southeast Asian development'. Power and Sidaway (2004) provide an overview of early texts written about the region that extend back to French, Dutch, German and British colonialists and explorers. While these texts were predominantly descriptive this approach changed in the 1960s when Buchanan wrote a landmark publication focused on the 'trends and forces that have shaped and are continuing to shape the turbulent and diverse nations of the region' (1967: 11, cited in Power and Sidaway 2004: 589). Buchanan's work signified the turning point for studies of the region, which began to analyse the region from the perspective of development theories such as modernisation and neo-Marxism. Since then there has been a large number of books that have adopted various economic and non-economic perspectives to explore regional development (e.g. Barlow 2000; Beeson 2004; Chia 2003; Dewitt and Hernandez 2003; Dutt 1996; Hill H. 2002; Hill R. 2002; Leinbach and Ulack 2000; Mackerras 1995; Rigg 1994, 2003; Rodan *et al.* 2001; Savage *et al.* 1998; Sien 2003; Weightman 2002; Yah 2004).

The aim of this book is to build on what has gone before to provide an introductory text that provides an updated account of development and development research within Southeast Asia. It is pitched at people who are new to the region or new to the subject of development and are interested in a challenging thematic overview of the region. In this sense the book is written to complement rather than compete with other academic texts by providing an easily accessible but theoretically engaging introduction to development in Southeast Asia. In addition the book adopts a unique perspective based on equitable development principles by arguing that development should improve the quality of people's lives by ensuring that access to the opportunities and benefits of development are equally available to all. While such principles may be difficult or even impossible to realise in practice they do provide a lens for interpreting development, and assessing the success or otherwise of different development processes. Such a perspective is not controversial; instead it could be argued that all the development theories listed above incorporate the goal of equitable development in one form or another, it is just the process that differs. Many of the economic development theories, for example, focused upon economies because they believed a vibrant economy equated with social wellbeing and felt boosting GDP would ensure more money would be available for all. Similarly grassroots development theorists believe putting the power of decision-making into the hands of the community is the best way to ensure that the quality of life of those communities will improve. To secure truly equitable development, in which all parties benefit appropriately and are equally empowered in decision-making processes, may well be illusory, yet this should not prevent such noble goals from being pursued or from acting as guiding principles. This book charts the progress of different states in Southeast Asia towards more equitable forms of development by looking at how development is reshaping social and physical spaces. The aim is not to provide solutions to unjust development but instead to expose the breadth of challenges present in what is often referred to as a 'successful developing region'.

To meet these aims the book is organised around four conceptual themes that provide a framework for examining the region. These themes draw upon some of the more recent development theories that have evolved since the concept of development was broadened to incorporate non-economic goals. They are:

- *Equality and inequality*: the comparative access people have to wealth, goods and services between and within national borders.
- *Freedom and opportunity*: the degree of autonomy people have to influence political processes and pursue their own visions of the future.
- *Participation and empowerment*: the relative involvement of communities in shaping and implementing development practices.
- *Environmental sustainability*: the long-term environmental impacts of current development strategies.

These four themes form a lens through which Southeast Asian development is interpreted and analysed in the chapters that follow.

Structure of the book

The remainder of the book is divided into eight chapters that thematically address different aspects of Southeast Asian development. Chapter 2 provides a historical context by outlining pre-colonial and colonial empires and looks at how decisions made in these periods are still shaping development today. Chapters 3–5 concentrate upon the economic, political and social dimensions of development. Chapter 3 examines the economic unevenness that has resulted within and between market-led and state-led economies and outlines ongoing challenges posed by the Asian economic crisis and growth based on FDI. Chapter 4 looks at the progress of the region from the perspective of good governance. Special attention is devoted to decentralisation programmes and the ramifications of these on political freedoms and local empowerment. Chapter 5 explores social transformations and the participatory potential of civil society and social movements to influence development. Case studies focus upon class, religion and gender. The subsequent three chapters examine development in urban, rural and natural spaces. The megacities of Southeast Asia are examined in Chapter 6 with a particular focus upon the housing and employment challenges facing urban authorities as rural people continue to migrate to the cities. Chapter 7 explores why people are leaving the countryside as well as the agrarian and non-farm transformations reshaping rural spaces. Chapter 8 analyses the impacts development is having upon natural spaces, and the risks natural spaces, in the form of hazards such as the Asian tsunami, pose to human development. The final chapter concludes by returning to the themes of the book to report on the achievements and ongoing challenges facing Southeast Asia from the perspective of equitable development.

Summary

- **Southeast Asia is extraordinarily diverse in terms of economies, politics, demography and culture.**
- **It is often considered to be a 'successful developing region' because of its overall economic progress, however there is considerable diversity between and within countries.**
- **The definition and theoretical construction of development is contested and is changing from a focus on macro-scale economic indicators to a range of more local social, cultural and political indicators.**

- **The book adopts the principles of equitable development to analyse Southeast Asian development by focusing upon equality, political freedoms and opportunities, participation and empowerment, and environmental sustainability.**

Discussion questions

1. Discuss how and why development theories have changed.
2. Explain what you think are the most important aspects of development and why.
3. Discuss the types of local, national, and global institutions that are likely to be empowered and disempowered in each of the development theories discussed in this chapter.

Further reading

For an excellent introduction to a range of development issues see:
Potter, R., Binns, T., Elliot, J. and Smith, D. (2004) *Geographies of Development*. Harlow: Pearson/Prentice Hall.

For good general overviews of development theories see:
Willis, K. (2005) *Theories and Practices of Development*. London: Routledge.
Todaro, M. and Smith, S. (2006) *Economic Development*. Boston: Pearson.

For an introduction to influential development thinkers see:
Simon, D. (ed.) (2006) *Fifty Key Thinkers on Development*. London: Routledge.

For an introduction to grassroots development thinking see:
Chambers, R. (1997) *Whose Reality Counts? Putting the First Last*. London: Intermediate Technology Publications.

For introductions to post-development thinking see:
Escobar, A. (1995) *Encountering Development: The Making and Unmaking of the Third World*. Princeton, NJ: Princeton University Press.
Power, M. (2003) *Rethinking Development Geographies*. London: Routledge.

Useful websites

The United Nations Human Development Programme website and reports. These include the annual Human Development Reports which allow you to compare countries according to various indicators: http://hdr.undp.org/reports/
The World Bank website provides insights into this powerful institution's perceptions of development: www.worldbank.org/
In contrast Oxfam, an influential international NGO, provides its own grassroots perspectives of development: www.oxfam.org/

2 Setting the scene for development: pre-colonial and colonial Southeast Asia

Introduction

Development, in its most recent incarnation, began spreading to Southeast Asia in the wake of the Second World War. Prior to this period, however, Southeast Asia had a long history of civilisations, kingdoms and communities that were evolving in response to internal and external forces. This chapter explores the evolution of Southeast Asia up until the independence period and looks at the types of development challenges independent states inherited from previous eras. Attention is directed to indigenous empires and their early contact with foreign powers, the influence of pre-colonial European contact upon international trade routes through the region, and the later more formal period of European colonialism. It was during these periods that the territorial boundaries of modern Southeast Asian states were delineated, their positions within global economies confirmed, and the early roots of modern geographic and social inequality were emerging. Through these periods the transformation of Southeast Asian societies can be observed, from their indigenous values and styles of governance to modern nation-states with apparatuses and imaginaries oriented towards the pursuit, promises and dreams of development.

Pre-colonial empires and influences

Southeast Asia has a long and complex history of competing kingdoms and empires that pre-existed modern development. Some of the most powerful and influential early kingdoms included the Funan and Khmer Empires based in contemporary Cambodia, Dai Viet and Champa in what is now southern Vietnam, Pagan in present-day Myanmar, and the maritime trade empires of Srivijaya based in Sumatra and the Sailendras of central Java (see Figure 2.1).

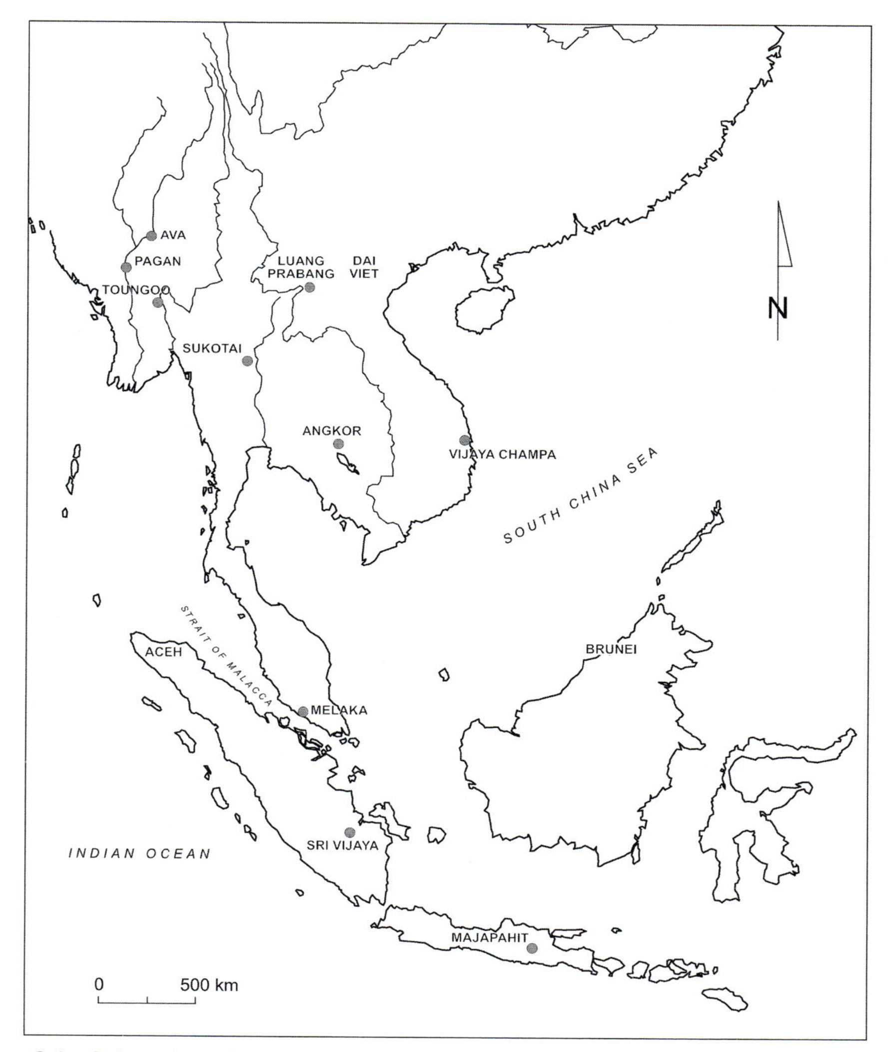

Figure 2.1 Selected empires of pre-colonial Southeast Asia.

At no point was all of what we now refer to as Southeast Asia ever united as a single empire, instead smaller kingdoms rose and fell as power shifted according to conflicts, alliances, population movements and trading relationships. Many of these early empires formed the spatial blueprint for modern-day territorial boundaries, however it is worth noting that power was conceptualised quite differently during these early periods. Instead of being absolute within national borders, as it is enacted today, power was seen to permeate outwards from the centre of the kingdom in what is known as a *mandala* system, becoming weaker with distance from the core. The centre of political and religious power lay at the centre of the *mandala* while subsidiary political units, such as principalities, states and sultanates, would have varying degrees of political, economic and spiritual freedom depending on their (in)accessibility to the centre. Those at the centre extracted resources from surrounding areas to build their wealth, providing an early pre-colonial example of uneven development. In contrast communities in hard to reach places beyond the *mandala*, particularly in the forested highlands of mainland Southeast Asia, evolved with relatively little contact with the outside world. Many of these groups still live outside mainstream society, residing in highland forests and boasting ethnically unique and distinct lifestyles, languages, customs and patterns of dress.

The most important external forces during this early period came from the powerful neighbouring civilisations of China and India. China was particularly influential in modern-day Vietnam, which was administered as a Chinese province until independence in AD 939. Vietnam inherited Confucianism and sophisticated court and governance systems from its neighbour, with Dai Viet being the only Southeast Asia empire that claimed spatial boundaries to power rather than *mandala*-style systems (see Osbourne 2000). While China occasionally constituted a military threat to the region its main influence was in terms of trade. Its flourishing empire was enmeshed in trade with Europe, India and the Middle East making Southeast Asian ports, particularly those of the Srivijayan Empire, important early nodes in the East–West trade routes. India also influenced Southeast Asia through trade and the travels of Brahmin priests. The priests taught Hinduism and Buddhism and eventually contributed to the conversion of most easily accessible parts of the region, with the exception of the Philippines, to these religions. Hinduism has since fallen from favour but Buddhism remains the dominant religion in mainland Southeast Asia affecting regional interpretations of development (see, for example, the Thai king's Buddhist vision of a sufficiency economy discussed in Chapter 5).

International trade accelerated from the fifteenth century onwards (see Reid 1992). Improvements in shipping technology saw more boats being built that were able to travel longer distances and carry heavier loads. Ports began to spring up along the coastlines of Southeast Asia with some becoming among the most important in the world. Melaka (Malacca), for example, which was founded on the west coast of the Malay peninsula in the fifteenth century, became one of the most important stopping points for the India–China trade and grew to a population of over 100,000 people, a significant number by world standards at this time. Melaka's early success owed much to its geographical location but also to the decision of its founder to convert to Islam, a relatively unfamiliar religion within the region. This heightened the attractiveness of Melaka to Islamic traders from the Middle East and India who preferred it to competing ports. Islam soon spread throughout Malaysia, Indonesia and Brunei, with Banda Aceh, on the northern tip of Sumatra, becoming known in the Islamic world as the 'Veranda of Mecca'. Islam is becoming increasingly political in these archipelagic states where it is being used to question the appropriateness of economic development directions (see Chapter 5).

The flourishing of important port cities sprinkled across the region kept Southeast Asia abreast of technological advancements and enmeshed in world economies. Local export industries grew as markets for items such as pepper, cloves, cotton, sugar and benzoin (used in incense) were found as far afield as Europe, Persia, China and Japan. Such demands began to alter the rural landscapes near port cities as communities began producing crops for export rather than local consumption, moving from subsistence lifestyles to ones based on trade and commerce. Port cities also began importing goods, with local markets showing a particular preference for the bright colours and fine weave of Indian cloth. The benefits of this period, however, did not flow evenly throughout the region. Instead most benefits were captured by coastal areas that had access to international trade routes as well as any core cities to which coastal areas were *mandala*-style subsidiaries. More distant areas were excluded from such trade due to undeveloped transport infrastructure that prevented the efficient transfer of goods. The booming economies of the ports compared to non-trading areas would become enhanced in the colonial period providing some of the early foundations for uneven development in Southeast Asia today.

Early European intervention

Southeast Asian ports were becoming important hives of commerce with traders from around the world valuing their location between China and the Middle East. However this was to change when European maritime empires from Portugal, Holland and Spain, used force to control trade coming in and out of the region. The thriving metropolis of Melaka was one of the first targets, being highly valued for its strategic potential. Reid (1992: 488) quotes the opinion of one early European: 'The trade and commerce between the different nations for a thousand leagues on every hand must come to Malacca . . . Whoever is lord of Malacca has his hands on the throat of Venice.' Portugal launched a bloody invasion which costs hundreds if not thousands of lives to eventually capture Melaka in 1511 and a number of other port cities soon after. The Portuguese asserted their authority through a style known as direct rule where they curtailed the political freedoms of the local populations and ruled directly by force. This allowed them to profit from taxes applied to traders using the port while subduing any local protests. Portugese rule was harsh and unpopular and they were eventually forced out of all their port cities, apart from Dili in Timor-Leste, by a superior Dutch fleet. As will be explained in Chapter 4, it is precisely because the Portuguese stayed in Timor-Leste that it stands as a sovereign nation, as opposed to an Indonesian province, today.

Holland played a vital role in shaping the boundaries of modern-day Indonesia. Prior to their arrival Indonesia was a diverse and sprawling mass of principalities and sultanates incorporating diverse communities with multiple languages, cultures and economies. Indigenous empires had come and gone, none even attempting to unite the entire archipelago into a single unit. Through the United East India Company (VOC) the Dutch managed to do just that, by declaring what is modern-day Indonesia the Netherlands East Indies and controlling local trade routes. The powerful VOC navy forced merchant and other ships passing through its waters to pay various commissions or risk being attacked. The Dutch set up forts similar to the Portuguese but also governed through indirect rule by entering into agreements with local rulers that maintained their sovereignty as long as they agreed to only trade with the VOC at VOC-determined prices. In some places they requested that particular crops would be grown that would they would then sell on to international markets for significant profits. Coastal economies slowed during this period as ports and traders received less value for the goods they were producing, and had to pay more to access goods from other parts of the world.

A similar process took place in the Philippines when the Spanish arrived and conquered Cebu in 1565. However lying east of mainland Southeast Asia the Philippines was not as ingrained in the East–West trade routes and the Spanish push appears to be driven as much by their desire to convert Filipinos to Catholicism as it was to profit from Filipino economies. The Spanish succeeded in both, only experiencing resistance in the southern islands where people had already converted to Islam as a result of contact with Middle Eastern traders. The Catholic/Islamic tensions that arose during this period continue through to the present day as successive administrations have struggled to cater for southern demands for political autonomy, or to provide a more equitable distribution of wealth and services to these southern islands. However the Spanish were more interested in investing resources into the country than the Dutch or Portuguese. They built the oldest European college in Southeast Asia, the college of Saint Thomas in 1611 and the Royal Philippine Company, modelled on the VOC, was mandated to invest 4 per cent of its profits in economic development schemes such as sugar plantations in the archipelago (Owen 2005). In many ways Spain was the earliest European power to adopt a colonial model of governance in Southeast Asia.

This early period of European intervention saw Southeast Asia slowly become secluded from trade and technological innovations taking place in other parts of the world. As Reid (1992: 504) writes:

> In order to preserve what they could of cherished values, comfortable lifestyles and familiar hierarchies, Southeast Asian states had disengaged from an intimate encounter with world commerce, and the technology and mind-set that went with it . . . The states that now had to confront the new Western onslaught now lacked the technology, the capital, the bureaucratic method, and the national coherence of their opponents, and would be made to pay dearly for it.

The slow down in international trade only affected coastal areas directly but had ripple effects through the broader region. The new technologies, ideas and goods that would have otherwise permeated Southeast Asian society were kept at bay and the potential benefits of trade were not realised. As a result the well-established empires on mainland Southeast Asia, which were particularly strong in areas of

Myanmar, Thailand and Vietnam, were not as technologically developed as they might otherwise have been. Secluded from international trade Southeast Asia began falling behind in terms of global technological developments weakening its ability to resist the next more comprehensive wave of European intervention.

European colonialism

The Industrial Revolution that gripped Europe in the early nineteenth century heralded a shift away from economies based on largely rural manual labour, similar to those in Southeast Asia, to ones dominated by machines and factory labour. Industrial innovations, particularly the invention of the steam engine, which could be used for pumps in mining, shipping or locomotives, as well as in industrial manufacturing, saw an increasing demand for raw materials as well as foodstuffs to feed factory workers. These demands drove a new wave of European colonialism which, assisted by superior weaponry and transport technology, soon spread throughout the world. Southeast Asia was a key focus, particularly after the construction of the Suez Canal that dramatically reduced shipping distances to Europe. This new phase of European contact sought to not only control maritime trade routes, but also to control whole countries and reorient them towards the needs of industrialising Europe.

The key colonising powers were the British in Burma and Malaya, the French in Indochina (Vietnam, Laos and Cambodia), the Dutch deepened their control within the Netherlands East Indies, the Portuguese retained Portuguese Timor and the Spanish ruled the Philippines until forced to cede control to the US in 1899 following the Spanish–American War (see Figure 2.2). The relationship between colonisers and the colonised reflected the differing approaches of different colonial nations, and sometimes of individuals within the colonising force, as well as the reactions and resistance of Southeast Asian societies. The powerful empires of Burma and Dai Viet, for example, mounted considerable resistance to the British and French respectively, with colonising powers only succeeded in defeating the ruling sovereigns, who were lagging behind advancements in weapons technology, after series of wars that stretched over periods of 61 and 27 years respectively. Colonial rule in such places was tightly controlled and often brutal in order to repress any further thoughts of rebellion among the defeated populations. Elsewhere, particularly in the

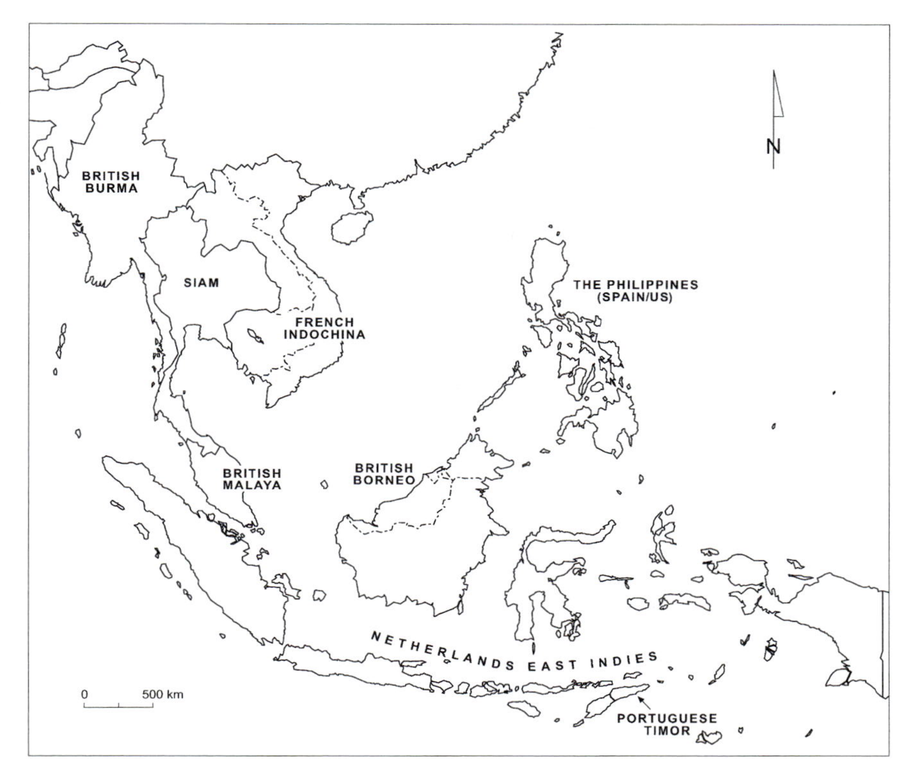

Figure 2.2 Colonial empires of Southeast Asia (late nineteenth century).

peninsula where trading outposts had already been established and power was scattered among much smaller geographical units, colonial opposition was less unified. In Malaysia, for example, the British came to power, initially through taking administrative control of Malacca, Penang and Singapore, known as the 'Straits Settlements', and later through extending trading relationships and forms of indirect rule through the interior of the Malay Peninsula. The Netherlands, likewise, gradually extended their authority through Indonesia, sometimes peacefully through clever politics and negotiations, but at other times, in places such as Bali and Aceh, brutally through long wars of violence and aggression. Siam, the predecessor to modern-day Thailand, was the only kingdom to escape colonisation; an outcome of clever negotiations on the part of the Siamese kings who played on French and British reluctance to share what would be a hotly contested colonial border.

The extent of change that took place across Southeast Asia during the colonial period cannot be underestimated. New governance systems were implemented, new economies evolved and societies were fundamentally transformed under the hands of the colonialists. From the perspective of equitable development both positive and negative changes can be observed. The region, for example, gained new technologies, improved farming outputs and experienced long periods of comparative peace leading to significant increases in population. On the other hand if political freedom is a core quality of equitable development there was precious little of this; democratic processes were largely absent and political opponents imprisoned if they spoke out about independent futures. Some colonialists saw the suppression of political rights as a necessary sacrifice if other social and economic advancements were to be made, an interesting forerunner to the Asian values argument that is discussed in Chapter 4. It was a time of mixed fortunes in which local landscapes become reoriented towards the production of goods for European profit. A cultural transformation also took place when relatively independent Southeast Asian societies became exposed to desirable European technologies, goods and luxuries. From a post-development perspective it was when the seeds of development's desirability in the post-colonial period would first be sown.

Political transformations

Colonialism initiated the transition from *mandala* governance systems to territorial conceptions of authority where power is constant within geographically defined spaces. In other words the power and influence of a central authority is just as strong at the edges of its territory as it is at its core. Colonialism was accompanied by new mapping technologies that outlined European claims to territory, with such claims forming the territory borders of Southeast Asian states today. In establishing and agreeing upon borders the risks of conflict between colonial powers was minimised and functional territorial governance systems could be constructed. Many of the borders were mapped with regard to existing polities, particularly on the mainland with British Burma, independent Siam, and the French territories of Vietnam and Cambodia roughly matching the claims of pre-existing kingdoms. In other places such as the Spanish Philippines and the Netherlands East Indies new states were formed which lacked any pre-existing indigenous political unit. Strangest of all, perhaps, was the British decision to combine parts of northern Borneo with Malaya as the Federated Malay States, which was as much a response to fears that Germany might stake a claim to the region, than any well-established links between the Malays of the peninsula and those in Borneo (see Tarling 1992a).

Governance within these territorial boundaries oscillated between direct and indirect rule, reflecting the ideals of the colonialists, the existing political structures of the local populations, and the geographic terrain colonialists were attempting to control. Forms of direct rule were employed in Burma, Vietnam, parts of Indonesia and the Philippines, while the British indirectly ruled Malaya by appointing advisors who, while officially under the control of local sovereigns, gradually came to dominate local politics and decision-making. What was constant, however, was the attempt to gradually assume centralised control in which uniform laws and taxes could be applied equally to all those within territorial borders and could be enforced through police and judicial systems. The autonomy of distant polities to effect their own laws and taxes was gradually erased as colonial governments grew in power and influence. In reality, however, no governance system was ever absolute and accepted, instead anti-colonial movements and peasant uprisings occurred within most colonial states and informal power structures survived, albeit in modified forms (see Christie 1996). For example, powerful interpersonal patronage networks in which influential members of a village or region would look after the interests of others in return for their political or economic support, continued during the colonial period and continue to exist today. In other words there were dualistic authority systems at work, the formal authority of the colonialists, and the informal networks of the indigenous populations.

The declaration of territorial laws, the training of police and introduction of judicial systems and processes of punishment were designed to further European control of colonial populations. The power of previous sovereigns and authorities, particularly religious authorities, was gradually usurped as uniform laws and processes were introduced. Power was centralised, and to a point, depersonalised through written laws. This secularisation and centralisation of authority was inherited by post-colonial governments with important ramifications for development. On the positive side it provided post-colonial rulers with accepted systems of governance, law and justice through which leaders could promote harmony and pursue development. This was particularly important for early development theories such as modernisation and neo-Marxist theories that believe the central authorities are crucial for development. Later development theories from neo-liberalism through to post-development, however, have tended to portray centralised governance as a hindrance to development. Many countries in Southeast Asia are now experimenting with decentralisation programmes to overcome the inefficiencies and problems of central government.

Of particular relevance to more recent conceptualisations of development that incorporate concepts of political freedom, participation and empowerment was the distribution of ethnic groups within borders and their subsequent relationships with centralised states. In many cases ethnic groups that had until then led relatively autonomous lifestyles became integrated into much larger political units. This is the case in the Netherlands East Indies, for example, where remote communities from Sulawesi or Sumatra became incorporated into the Dutch Empire that was locally administered from Java. Similarly many highland groups in mainland Southeast Asia become incorporated into lowland empires with which they had few relationships or influence. Their autonomy has been threatened ever since, with states claiming ownership of their land and introducing policies of assimilation and integration (see Box 2.1). Refusal of ethnic minorities to cede to centralised authority has contributed to long wars of independence in parts of Myanmar, Indonesia and the Philippines, posing considerable challenges to free and peaceful development. In other cases colonial borders have divided major ethnic groups so that the majority of the population become incorporated into territories controlled by people of a different ethnicity. For example 19 million ethnic Laotians live within the Thai border as opposed to just 3.7 million in Laos, and far more ethnic Malays live on the Indonesian island of Sumatra than in Malaysia (Cribb and Narangoa 2004). In centralising authority and spatialising power the ability of ethnic groups to pursue their own visions of the future has been severely eroded.

Economic transformations

The economies of Southeast Asia, like the political systems, underwent fundamental change during the colonial period. What had previously been relatively insular undeveloped economies, particularly since Europe had taken control of maritime trade routes, were turned into monetised economies oriented around the export of primary goods to Europe. There was a rapid increase in acreage devoted to export crops, a proliferation of mines extracting valuable minerals, widespread harvesting of hardwood forests and a substantial extension

Box 2.1

Drawing borders, creating minorities

The territorial borders that were drawn during the colonial period have caused ongoing problems for the dozens of different ethnic groups living in the highlands of Southeast Asia. Millions of people lost their independence and became minorities within states dominated by lowland groups with whom they had little knowledge or affinity. Subsequent minority–majority relations have been strained with mistrust, misunderstandings and mutual antipathy characterising interactions. One group that has suffered from the establishment of territorial borders has been the Hmong minority group that, despite an overall population of 4–5 million people, has been unable to negotiate an independent state of its own. Instead its people been divided between China (estimated at 3 million), Vietnam (670,000), Laos (230,000) and Thailand (130,000). The Hmong, like other highland minority groups in Southeast Asia, have developed their own language, customs and style of dress, which are distinct and foreign to lowland peoples, making them subjects of state suspicion and persecution. This has included criminalising their traditional agricultural systems and claiming the forested land on which they live as state property. Laotian Hmong have suffered more than most for their role in supporting the Americans when the communist Pathet Lao took power during the American/Vietnam War. Some 300,000 Hmong fled to camps in Thailand and have since been granted refugee status in the US (275,000), France (15,000) and elsewhere.

Another ethnic group divided that has suffered as a response of colonial borders has been the Karen people who are divided between Myanmar (4 million) and Thailand (400,000). During the colonial period the Karen, and many other highland minority groups in Myanmar, were ruled indirectly and resented the shift to direct lowland rule that accompanied Burmese independence. They have since fought for independence and autonomy from the central state but have been the subject of brutal 'Burmanisation' campaigns launched by the Myanmar military, which have included arbitrary executions, rape, forced labour and forced resettlement. Thousands of Karen people have fled Myanmar to the security of border camps in Thailand where their movements and lifestyles are restricted by their lack of citizenship. Those who risk leaving the camps are forced to work illegally in low paying jobs in construction or sex industries and must pay bribes to authorities or risk repatriation to Myanmar should they be discovered. While there has been some positive progress in minority–majority relations in Southeast Asia, what Clarke (2001) calls a shift from 'ethnocide to ethnodevelopment', there is a long way to go before minority groups will have reason to cease ruing the territorial maps drawn by colonialists.

Sources and further reading: Hamilton-Merrit (1994), Lemonie (2005), Clarke (2001), Grundy-Warr and Yin (2002) and improvement of transport infrastructure. These changes had both positive and negative ramifications for Southeast Asian societies and their subsequent development. Economic change was propelled by colonial investment, encouragement and coercion, as well as indigenous and migratory entrepreneurship. Some colonies, such as the Netherlands East Indies, relied on coercive mechanisms such as the 'culture system' that required a land tax to be paid in cash to colonial authorities. This forced subsistence farmers, who had previously relied on barter and non-monetary exchange systems, to grow cash crops that could then be sold to the VOC at fixed low prices so that they could raise the capital to pay the tax. Exports boomed as a result of this tax but Indonesian families suffered as they struggled to feed themselves from the small portion of land still devoted to subsistence rice production. Less oppressive measures were adopted by the British in lower Burma who lowered taxes, provided secure land titles to peasants, improved irrigation and transport infrastructure and provided milling and export facilities. In doing so they transformed the subsistence-style economy into one of the largest rice exporting regions in the world. The French achieved a similar outcome in southern Vietnam by expanding the indigenous canal system (partly through forced labour), improving transport, irrigation and, ultimately, the export efficiency of the region (see Elson 1992). Other areas, such as the west coast of Malaya and parts of the Philippines, became famous for the establishment of large non-indigenous crop plantations such as rubber, sugar, tobacco, coffee and copra. Malaya became the world's biggest producer of rubber while the Philippines became a key player in the sugar industry. Non-farm pursuits included timber harvesting, which took place under the management of newly formed colonial forestry agencies (see Bryant 1997 for a Burmese example) and tin mining which, while established by Chinese migrants in the pre-colonial era, increased in size and scale with the introduction of new extractive technologies. Southeast Asian landscapes, even those of Thailand, which sought to capture overseas markets with British assistance and advice, were fundamentally reoriented to the production of materials for European industrialisation.

Accompanying the booming export industries was a formalisation of financial systems. None of the pre-colonial economies had a recognised currency, instead port cities traded in Mexican dollars and coins from around the world, as well as traditional trading items such as gold, silver and, in Burma and Thailand, cowrie shells. Much trading was based on barter and exchange with buyers often knowing one another and basing trades on trust and promises of outputs from upcoming harvests. This changed under colonialism with new local currencies being introduced in the early 1900s enabling new occupations and encouraging the further commercialisation of agriculture. Monetisation saw European and Asian banks evolve in urban centres and formalised the position of the region within the world economy. However it also changed the way Southeast Asians interacted with their local landscapes. No longer was the local environment seen primarily as something to support the local community, and therefore something to be nurtured and cared for. Instead it became embedded with economic values, something that could be exploited for monetary gain. Similarly personalised trading relationships became less important than finding the best possible price. A cultural change was taking place.

The export economies of Southeast Asia were based on a 'vent-for-surplus' model in which international trade provided the 'vent' or outlet for surplus resources that would not otherwise be produced due to a lack of local markets (Huff 2003). In most parts of the region this meant surplus land and labour, which included labourers from South Asia, was utilised in the service of overseas markets. Consequently there was little need for financial investment in capital-intense industries such as manufacturing. The resultant international economy was one based on the export of primary products and the import of manufactured products from Europe. It is here that the roots of a world trading system of the core–periphery type modelled in neo-Marxist theories can be found. Low value agricultural products were exported to Europe and high value manufactured goods, such as steel for railways or garments and textiles, were imported back to the colony. This trade was done in preferential ways during the colonial period as core countries had an interest in ensuring peripheral efficiencies, but nevertheless the lack of domestic industrialisation and diversification is startling, and something post-colonial governments would later have the challenge of overcoming.

A number of other economic development challenges also arose out of this period, three of the most significant being uneven development, dual economies and land ownership. Uneven development occurred due to the rapid economic growth of port cities and their hinterlands and areas connected to transport infrastructure such as roads and railways, as opposed to more remote areas that remained relatively untouched by economic advancements. Overcoming such unevenness has become a constant challenge for development ever since. Dual economies refer to the co-existence of formal monetised economies and the informal systems of exchange that continued to take place. Informal economies dominated more rural and remote areas but also play, and continue to play, vital roles in cities, particularly among the urban poor. Informal economies are untaxed and unregulated, something that will be discussed in Chapter 6, and often out of sight of colonial and post-colonial governments – preventing investment and improvements. Finally land was vital to vent-for-surplus economies and land laws were introduced that were heavily weighted in favour of colonial governments as opposed to traditional occupiers. As there were few pre-colonial land records colonial powers claimed sovereignty over land unless certain occupancy conditions were met. Land was then mapped and titled and distributed to 'friends' of the colonial powers, suddenly transforming the communities that had worked the land for generations into tenants. This land grab was particularly severe in the Philippines where successive governments have attempted, but mostly failed, to implement land reform (see Chapter 7). It has also had important impacts on forest dwellers whose homelands have been declared state land, transforming them into illegal squatters (see Chapter 8). Overcoming these and other negative impacts of colonialism has been central to many current development programmes.

Social transformations

During the colonial period Southeast Asian societies became much more diverse and stratified along race and class lines. When Europeans first arrived many lives were lost to illnesses and diseases for which Southeast Asians had few natural antibodies. After this time, however, population numbers were to rise significantly, reflecting a period of relative peace and harmony, an expansion of farmland and irrigation as well as an influx of migrants, some from Europe but the majority from the South Asia and China. Britain was involved in the recruitment of some 3 million Indian labourers coming to Malaya in the early twentieth century and 2.6 million to Burma from 1852–1937 (Elson 1992: 163). Some debate exists in regards to the degree of freedom these labourers had when coming to Southeast Asia. A neo-Marxist reading focuses upon indentured workers who they argue were virtual slaves, having to work off their transport and accommodation debts by working for low wages in difficult conditions on Malayan and Burmese plantations. In contrast reformist readings concentrate upon the higher incomes migrants could earn compared to Indian economies as well as the luxury they had in leaving their caste status 'at home', arguing they chose to move out of rational free choice. As Satyanarayana (2002) suggests both readings are likely to be correct but relevant for different groups of workers, such was the magnitude and diversity of the labour flows.

Most migrant workers intended to return home but many stayed on, particularly those who managed to secure high ranking bureaucratic positions, often forming a racial buffer between the European colonials and indigenous elites. Millions of Chinese also migrated to the region, some indentured and some free, to work in plantations and mines as well as set up their own entrepreneurial businesses in expanding economies. The most obvious influx of Chinese was into the new city of Singapore, which grew from its modest establishment population in 1819 to 52,000 citizens by the mid-1840s, 32,000 of whom were Chinese (Osbourne 2000: 103). Chinese migrants successfully sought out new business opportunities during the colonial period and have since become an economically powerful ethnic group in most Southeast Asian economies. Their disproportionately high representation in upper income categories in most countries has been a source of frustration for lower income groups and a challenge for those concerned with distributing the benefits of development evenly throughout society (see Chapter 5). Malaysia has been the most active and controversial in this regard having designed race-based policies that promote the economic advancement of Malays over other ethnic groups (see Box 2.2).

Box 2.2

Rubber, tin, multiculturalism and bumiputra *in post-colonial Malaysia*

The introduction of rubber production and tin mining not only shaped the expansion and modernisation of Malaya's colonial economy, but has also had lasting impacts on the country's demographic profile and related post-colonial development policies. Much of Malaya's early colonial success was derived from low cost labour sourced from South Asia, particularly India, as well as China, which kept production costs down within rubber plantations and tin mines. By the end of the Second World War there were approximately 850,000 Indians in Malaya and around 1 million Chinese. Both groups, but particularly the Chinese, prospered during colonial times forming good relationships with the British and finding employment in profitable industries.

Concern about the influence of immigrant groups upon post-colonial development affected the writing of a Malaysia's constitution in 1957. An 'ethnic bargain' was arranged in which immigrant groups gained citizenship on the provision that ethnic Malays, referred to as *bumiputra* or 'sons of the soil', were granted special privileges as indigenous people. These early racial rights were extended after anti-Chinese riots caused at least 190 deaths and destroyed thousands of homes during May 1969. The government used the riots to introduce a New Economic Policy (NEP) in 1971 that further extended the rights of *bumiputra* so they received preferential treatment in areas such as scholarships, employment, trade licenses, housing and many other sectors of society. The NEP policy has proved extremely controversial, igniting heated debates about what it means to be Malaysian or Malay, not least from the marginalised *Orang Asli* communities who pre-date the arrival of Malays and threaten the base premise of *bumiputra* legitimacy. Nevertheless the policy has contributed to a reduction of poverty among Malay citizens and is still a prominent development policy today. The question is whether the benefits of these gains outweigh the negatives of reinforcing damaging ethnic stereotypes and ethnic segregation.

Sources and further reading: Lian (2006), Chakravarty and Roslan (2005), Tan (2001)

Social transformations also took place through the provision of colonial ideas and services into local communities. Many colonialists legitimised their presence in Southeast Asia by positioning themselves as trustees for their colonial subjects, believing they had a role to 'improve' existing societies through what the French referred to as the '*mission civilisatrice*', and the British as the 'white man's burden'. Such approaches tell much about the colonial mindset and their lack of understanding of existing indigenous societies, something that would be duplicated by modernisation theorists. As a result a range of important social programmes were introduced. One of these was the provision of new education systems based on European schooling models that fitted the civilising mission of the colonialists while they also contributed to producing a more effective and malleable labour force. Colonialists also invested in health systems and hospitals, introducing Western medicine to the region and providing the foundations for post-colonial health care. Combined with the introduction of impressive European technologies, such as steam-trains and automobiles, the provision of education and health services encouraged local interest in Western knowledges and worldviews. This was particularly the case among local elite families whose children were afforded further education in the home countries of the colonialists before coming back to take important posts in colonial and post-colonial regimes. From a post-development perspective it was during these colonial times that local elites began to nurture Western imaginaries and desires that taint indigenous visions of development. While such imaginaries were far from universal, indeed access to education and health services was extremely unbalanced – something post-colonial development programmes have tried to address – they did affect some of the most powerful and important local families; the ones that would lead the region and pursue development after independence.

The end of colonialism

Internal and external forces conspired to end Europe's reign in Southeast Asia soon after the Second World War. By this time the nature of anti-colonial protest had changed from its earlier incarnation, which sought a return to the 'old ways' and the kingdoms of the past. These protests were often fought, and lost, by using traditional weapons and calling on spiritual forces that could not compete with the guns of the Europeans (see Christie 1996). The seeds of more modern anti-colonial movements derived, somewhat ironically, from the Western-style schools the colonialists had set up, exposing European political ideologies and nationalist dreams of independence to idealistic students. Rather than return to the past, the new anti-colonial movements aimed to replace the colonial governing apparatus with progressive independent ones of their own. Substantial anti-colonial movements formed in Vietnam, the Netherlands East Indies and Burma, prior to the Second World War, however it was Filipino activists who initiated the first successful uprising in Southeast Asia by forcing, with US assistance, Spain out of the country in 1898. It was a hollow victory, however, with the US subsequently enforcing its own colonial administration through a brutal war costing hundreds of thousands of Filipino lives from 1899–1913.

The event that would eventually lead to Southeast Asian independence was the Second World War. Colonial powers were focused on Europe and their colonies were surprised and under-protected when Japan entered the war and began marching southwards. As France had previously surrendered to Germany there was very little resistance experienced by Japanese forces when they marched on Indochina and occupied French territories in 1940. On the night of 7 December 1941 Japan attacked colonial forces in Malaya, the Philippines and Hong

Kong, as well as Pearl Harbor in Hawaii, and bombed Thailand to begin a four-year war in Southeast Asia and the Pacific. The invasion was devastatingly successful; all colonial governments in the region were defeated within six months of the first assault. Many independence movements initially welcomed the Japanese troops and their motto, 'Asia for Asiatics'. However the war and occupation were brutal and caused widespread suffering throughout Southeast Asia. Manila, for example, was the second most war-damaged city in the world after Warsaw in Poland (see Owen 2005: 272). Thousands of lives were lost as countries became split between those fighting with the Japanese against those who sided with the colonial power. Economies collapsed, infrastructure was destroyed, blockades prevented the imports and exports of goods, and law and order became increasingly erratic and violent. For independence leaders it quickly became apparent that Japan was more interested in reorienting Southeast Asian industries to fit the interests of Tokyo rather than true Asian liberation. As Owen (2005: 273) has written: 'It was an era of hardship, cataclysmic change, and political turmoil, which transformed the Japanese dream of "a new dawn for Asia" into a Southeast Asian nightmare.'

When the war turned towards the Allies after the Battle of Midway in 1942, Japan empowered quasi-independent governments in the Philippines, Burma and Indonesia in an effort to garner domestic support to resist the Allies. This was not enough for most independence leaders who, having experienced the hardships of Japanese rule, sided with the Allies to begin forcing Japan out of the region. The eventual surrender of Japan in 1945 signalled the end of its brief but brutal sojourn into Southeast Asian politics, however its impacts were to last much longer. In the period of uncertainty before colonial powers were able to re-establish themselves nationalist movements were able to organise and garner popular support for independence. Post-war global geopolitics favoured independence over colonialism and with Europe weakened by war the dream of Southeast Asian independence was within grasp. For some, such as the Philippines, Burma and Malaysia, independence leaders managed to negotiate relatively peaceful transitions to independence and inherited the colonial governance apparatus intact. In contrast Indonesia had to fight a four-year war for independence against Holland, while Vietnam had to fight much longer wars against France and then the US before it was unified and independent. In these countries, and in Timor-Leste, which underwent a further round of colonisation under Indonesia, the

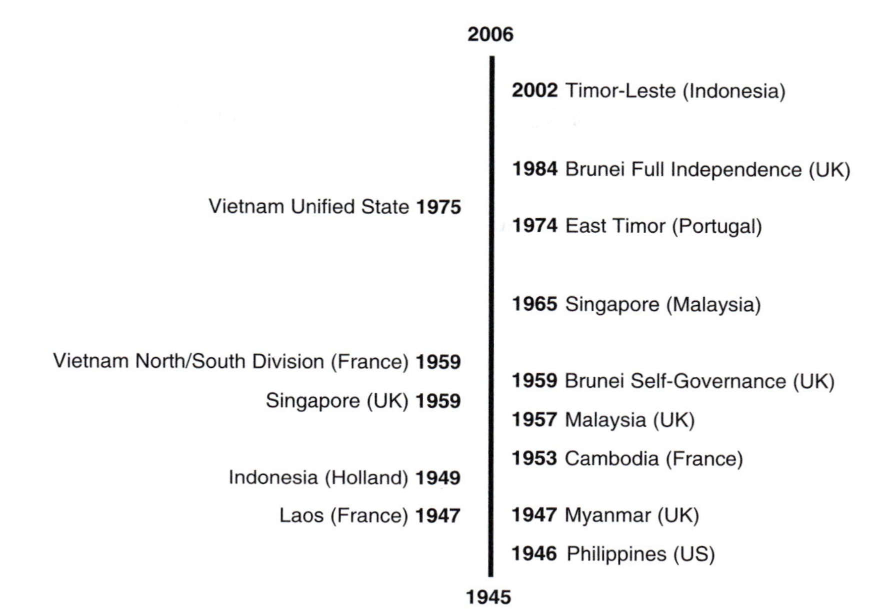

Figure 2.3 Timeline of Southeast Asian independence.

subsequent nationalist leaders inherited war-damaged states making future development challenges that much harder (see Figure 2.3).

Summary: progress prior to independence

A wide variety of Southeast Asian societies emerged out of the colonial period to grasp independence. Their diversity reflected unique pre-colonial indigenous societies as well as trade and interaction with some of the world's most powerful civilisations in India, China, the Middle East and Europe. On this diverse canvas the pursuit of more modern forms of development, the subject of the remainder of this book, was to occur. Before turning to post-colonial development, however, it is worth reviewing the strengths and weaknesses of pre-colonial and colonial progress from the perspective of equitable development.

Prior to European intervention there was considerable diversity in terms of wealth and opportunities within Southeast Asia. More powerful members of society lived in cities that were either oriented towards trade and commerce or governance and authority. Trade-based cities were exposed to the ideas, wealth, trends and technologies of passing traders while cities of governance were embedded with spiritual power, housing large palaces and religious monuments. The ancient civilisation of Angkor, for example, built the impressive highly detailed Angkor Wat, generally considered the largest religious monument in the world at the time. There was a diversity of opportunities available for people living in cities as they could pursue work as merchants, traders, builders, carpenters and a range of other urban-based services. Those in the rural areas, however, where the majority of the population lived, had little access to these urban opportunities, the majority of contact restricted to providing agricultural tributes to their rulers, or, in the case or war, providing their sons as soldiers. Osbourne (2000) has suggested that the majority of villages in pre-colonial Southeast Asia, while not entirely disconnected, were relatively closed to outside influences due to a lack of knowledge of the outside world and the undeveloped state of transport infrastructure. Villages were left to evolve semi-autonomously, particularly if they were at the outer edges of a *mandala*, and while secluded from technological advancements they were free to pursue their own visions of the future and evolve in sustainable and culturally relevant ways. However farming lifestyles were difficult and repetitive, opportunities for alternative lifestyles extremely limited and individual freedoms curtailed according to the cultural norms of that area. Progress, then, was uneven in pre-colonial times, with political autonomy and access to services and resources differing across space. The greatest strength of this era from an equitable development perspective was that Southeast Asian peoples were in control of their lands and free to develop in their own ways. This can be seen in the priority they gave to spiritual and religious development, with everyone from kings to peasants investing time and resources in religious teachings, offerings, buildings and infrastructure. Spirituality, something that is often overlooked in contemporary development, is still very important to Southeast Asia and can be seen in the numerous monuments and ceremonies celebrated throughout the region today.

Early European intervention slowed international trade and stunted growth in most countries within the region, with the possibly exception of the Philippines. In terms of political freedoms and opportunities it was a difficult time as early European maritime empires usurped the power of local sovereigns by restricting the autonomy and free trade of port towns. This curtailing of political autonomy was enhanced and extended with the onset of more comprehensive form of colonialism from the nineteenth century when decision-making became the domain of European authorities. The region underwent a rapid period of externally directed change in which entire economies, landscapes and societies became reoriented towards the needs of European industrialism. Some communities experienced considerable benefits during this period as food production increased, schools, hospitals and technologies spread, livelihood opportunities expanded, and wealth flowed into monetised economies. In some parts of Southeast Asia there was an increase in livelihood opportunities, particularly as transport infrastructure improved. However it was uneven development with access to goods and services dominated by newly emergent wealthy classes, local political freedoms and participation in decision-making restricted to small groups of colonial 'friends', and local cultural norms and values diluted by Western education and wealth. Progress was occurring, but it was a European vision of progress in which local participation, empowerment and political freedoms were subsumed beneath an unsustainable European grab for resources. Many of the advancements that had been made were damaged or destroyed during the Second World War and subsequent wars of independence.

Newly independent governments inherited substantial development challenges from their earlier experiences. On the positive side they inherited colonial governance systems and associated state apparatus such as police and judicial systems as well as basic education and medical systems. However these, as well as access to wealth, land and resources, were distributed unevenly generally favouring elite classes in large cities and port towns over people in more remote rural and landlocked places, including almost all of Laos. They also inherited territories that were comprised of multiple ethnic groups, many of whom were opposed to ceding power to ruling elites of different ethnicities. For newly independent countries such as Indonesia and Burma, the challenge of building singular national identities across diverse multi-ethnic terrains were to be considerable. Economically newly independent states inherited economic systems oriented around the export of primary products to Europe but very little local industrial activities. Economies were relatively unsophisticated, with expansive informal economies based on non-monetised transactions, and dependent on manufactured products from other places. Important challenges lay ahead in industrialising local economies and breaking free of the unprofitable and unsustainable trade relations that they had become locked into. Most importantly newly independent states inherited leaders who were usually Western educated and keen to pursue modernisation by engaging in development rather than exploring independent indigenous Southeast Asian alternatives.

Summary

- **Southeast Asia has a long history of pre-colonial kingdoms and empires, many of which have been economically and geopolitically significant because of their important location in East–West trade routes.**
- **Early European contact with the region suppressed development by controlling port cities and trade routes through the region.**
- **Colonialism rearranged Southeast Asian societies politically, economically and socially to fit the needs of European industrialisation and contributed to a reconceptualisation of governance and spatial power that provides the political foundations of Southeast Asian states today.**
- **Colonialism encouraged a social and cultural shift from the indigenous values and imaginaries of existing societies to more modern values and imaginaries favoured by development.**
- **The Second World War brought hardship to Southeast Asia but also provided opportunities for nationalist leaders to pursue strategies that would eventually lead to independence.**
- **The legacies of uneven development during pre-colonial and colonial periods continue to shape development patterns across the region today.**

Discussion questions

1. Contrast the strengths and weaknesses of pre-colonial and colonial eras from the perspective of equitable development.
2. Outline the multiple impacts Europe had upon Southeast Asia societies prior to independence.
3. Discuss the significance of changing spatial conceptualisations of territory and power.

Further reading

Many excellent books have been written on the history of Southeast Asia. For well written and easily accessible introductory texts see:

Osbourne, M. (2005) *Southeast Asia: an introductory history (9th Edition)*. Sydney: Allen and Unwin.

SarDesai, D. (1997) *Southeast Asia: past and present (4th Edition)*. Boulder, CO: Westview Press.

Other more detailed books include:

Tarling, N. (ed.) (1992b) *The Cambridge History of Southeast Asia Vol. 1: From Early Times to c. 1800*. Cambridge: Cambridge University Press.
Tarling, N. (ed.) (1992c) *The Cambridge History of Southeast Asia Vol. 2: The 19th and 20th Centuries*. Cambridge: Cambridge University Press.
Owen, N. (ed.) (2005) *The Emergence of Modern Southeast Asia: A New History*. Honolulu: University of Hawaii Press.
Steinberg, D. (ed.) (1987) *In Search of Southeast Asia: A Modern History*. Honolulu: University of Hawaii Press.
Hall, D. (1981) *A History of Southeast Asia*. London: Macmillan.

For an alternative history that focuses on unsuccessful nationalist struggles for independence see:

Christie, C. (1996) *A Modern History of Southeast Asia: Decolonisation, Nationalism and Separatism*. London: Tauris Academic Studies.

3 Economic development

Introduction

Improving economies by raising GDP has been the primary focus of development efforts since the end of the Second World War. Southeast Asia, as a region, is generally portrayed as being particularly successful in this regard, having moved beyond the primary industries it inherited from the colonial period to much more diversified economies that incorporate strong manufacturing and tertiary sectors. In the 1990s, for example, books were written with titles like *The Southeast Asian Economic Miracle* in which it was argued:

> There is widespread agreement that the world's most successful developing countries have been those of the Association of Southeast Asian Nations (ASEAN) and some of its East Asian neighbours in the 1980s . . . in any number of countries . . . one wishes the planners and bureaucrats might be able to visit Southeast Asia and copy some of the methods, technologies, and inspiration that have improved the lives of hundreds of millions of people while creating political stability and an expanding ability to share in the world economy.
>
> (Barber 1995: 243–245)

A 1993 World Bank report entitled *The East Asian Miracle* was similarly enthusiastic, attempting to identify the underlying causes of East Asian economic success and translate these as lessons for other developing economies (World Bank 1993). Not surprisingly, given the authoring institution's economic orientation, the report identified neo-liberal principles as the drivers of success, something that has became a source of much debate, as will be examined further in this chapter.

More recently, however, the plaudits for Southeast Asian economic development have softened. There are two reasons for this, the first being the Asian economic crisis that surprised everyone when it gripped the region in the late 1990s. The crisis showed how quickly economies linked to world markets and global investment patterns could collapse as well as the devastating consequences such disintegration could have upon people's lives and livelihoods. Millions of people lost their jobs and were plunged into poverty as businesses folded and the price of basic commodities skyrocketed. The shock was unexpected and deeply felt and while recovery has occurred, it has thrown a cloud over the region's future. The second reason why plaudits have softened reflects a greater appreciation of the importance of non-economic aspects of development. There are countless examples of GDP-raising activities that have negative social or environment impacts, many of which are explored in later chapters, which do not equate to 'positive change' for all. This chapter will overview the uneven economic development of the region and outline some of the challenges facing the future.

Economies of Southeast Asia

Southeast Asian economic landscapes are extraordinarily diverse. As a region Southeast Asia has well-established agricultural systems, extensive manufacturing industries, well-developed service sectors, including a booming tourist industry, and an emerging stake in high-tech knowledge industries. At the national scale Indonesia has the largest economy, reflecting its huge population, while Singapore, followed by Brunei, are the richest, both classified as high income economies by the World Bank (see Table 3.1). Singapore, however, has a highly advanced and diversified economy based predominantly on service and knowledge industries, particularly banking and finance, while Brunei's economy is undiversified, being largely dependent on the processing of its large but ultimately finite oil reserves (see Figure 3.1). Malaysia is classed as a middle income economy but has ambitious plans to reach 'developed nation' status by 2020 by investing in high-tech service and manufacturing sectors while retaining important links to the land

Table 3.1 ***Key economic development indicators***

Country	*HDI 2006 (ranking of 177)*	*GDP 2005 (US$ millions)*	*GDP per capita 2005 (US$)*	*Annual growth rate 2005 (%)*	*Population living beneath national poverty line (%)*
Singapore	25	116,711	26,821	6.4	n/a
Brunei	34	6,248	16,882	3.0	n/a
Malaysia	61	130,654	5,001	5.2	15.5
Thailand	74	176,559	2,726	4.5	13.1
Philippines	84	97,685	1,160	5.0	36.8
Indonesia	108	280,265	1,275	5.6	27.1
Vietnam	109	52,809	635	8.5	28.9
Cambodia	129	5,523	404	9.8	35.9
Myanmar	130	5,922	106	4.5	n/a
Laos	133	3,727	418	8.2	38.6
Timor-Leste	142	349	367	2.5	n/a

Sources: ASEAN Secretariat (2006), UNDP (2006), World Bank (2007)

Note: n/a – statistics not available.

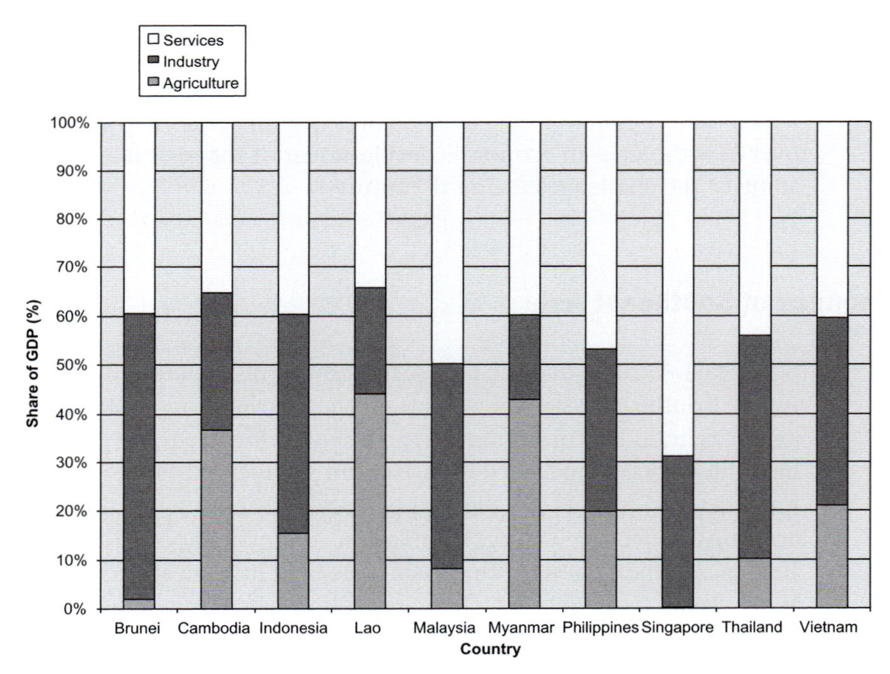

Figure 3.1 Share of GDP by major economic sector for ASEAN countries.

Source: ASEAN Secretariat (2005a)

through large agricultural plantations. Indonesia, Thailand and the Philippines are all grouped within the World Bank's lower middle income category and experience pockets of extraordinary wealth and extreme poverty. Finally Vietnam, Cambodia, Laos, Myanmar and Timor-Leste all fall within the World Bank's poorest category of low income countries where average per capita incomes come to less than US$3 a day. In these countries agriculture supports a much greater proportion of the population than the share of GDP it receives, indicating large pockets of rural poverty in each of these countries.

Hidden beneath the veneer of these formal statistics are the remnants of informal economies that pre-existed colonial eras and have adapted and modified according to changing social and economic circumstances. Informal economies are those that take place out of view of government, or in other words, they are untaxed and unregulated. They are usually small scale, often family run and reflect the entrepreneurial talents of those who are secluded from, or do not wish to participate within, formal economies. In the poorer countries of Southeast Asia informal economies can be of equal size and significance for the local population in terms of income, employment and the provision of services and goods as the formal sector. Examples in Southeast Asia include small scale food producers and street stall vendors, informal transport operators, labourers, fortune tellers, handicraft makers, hawkers, beggars and sex workers. Perspectives of informal economies have changed over time from early modernisation approaches that tried to either ignore or eradicate them, seeing them as inefficient and undesirable legacies of 'traditional societies', to more recent approaches that are more likely to recognise their value to society. Informal economies have been portrayed as neo-liberal exemplars of free market systems, popular manifestations of grassroots empowerment, and post-development celebrations of alternatives-to-development. No matter which perspective is adopted informal economies are of immense importance throughout Southeast Asia, particularly in those poorer countries where formal economies and employment opportunities are small.

The economic diversity of Southeast Asia reflects the different development paths newly independent states adopted upon independence. The more economically successful countries pursued capitalist market-based development strategies while Vietnam, Laos, Myanmar and Cambodia opted for socialist state-led approaches. These early post-colonial decisions have had long-term impacts for economic development patterns within the region.

Development of market-led economies

In the aftermath of the Second World War the global geopolitical environment changed from one dominated by colonial empires to a system shaped by the world's new superpowers, the US and the USSR. The political ramifications of this for Southeast Asia were significant, as will be discussed in the next chapter, however the economic consequences were just as important. Countries around the world were faced with a choice of siding with one of the two superpowers or pursuing a third less defined unaligned path. In Southeast Asia the new independent governments of Singapore, Malaysia, Thailand, the Philippines, Indonesia (the ASEAN-5) and eventually Brunei, chose to build on their colonial links by siding with the US. The decision to side with the West, despite Western empires dominating them during the colonial era, reflects the relatively peaceful transitions most of these countries made to independence. Indonesia is the exception as it had to fight for independence from Holland but it received important geopolitical support from the US, which formed the basis for a strong, ongoing and occasionally controversial, post-colonial alliance.

In siding with the West the ASEAN-5 countries committed to pursuing the capitalist market-led development models favoured by modernisation theory while rejecting, and often criminalising, the competing economic and political models associated with communism. The West provided aid, trade and financial support to assist these countries' progress through the stages outlined within modernisation to help them pursue the type of consumer society epitomised by the US. In the early years this meant support for the types of projects that had been initiated during the colonial era such as improving infrastructure, establishing financial services, commercialising agriculture, harvesting raw materials and improving the governance systems of the state. As a consequence early post-colonial economies looked much like the colonial ones with export-based primary industries such as timber, tin, rubber, rice, sugar and coffee, providing the bulk of foreign income. The main difference was a change of personnel controlling the country rather than any fundamental reordering of economic systems. However these new leaders soon began exploring ways of developing manufacturing industries in an attempt to diversify and grow their economies.

Huff (2003) has argued that one of the reasons why there was not more investment in manufacturing in the colonial period was because of the surplus land, labour and resources available for the pursuit of primary industries. This meant there was little need for capital-intensive manufacturing investments as employment and income was readily available at little cost. While this suited colonial rulers, who were mainly interested in profitability for the colonising country, it was of concern for post-colonial states who wished to become economically comparable or competitive with Western nations. Neo-Marxist theories such as dependency theory and world systems theory helped articulate the dilemma facing Southeast Asian leaders: they were exporting low value primary goods to core countries while importing high value manufactured goods back into their own countries in a system that was becoming increasingly profitable for the core, and increasingly exploitative for the periphery. They couldn't withdraw from exploitative global economic systems as they would not have the capital or technology to build manufacturing plants, but nor could they build plants that were as technologically efficient and economically competitive as that which existed in core countries. Local authorities looked to neo-Marxist theorising that suggested the state had to become more actively involved in the economy, functions that are now associated with the idea of a *developmental state*. Developmental states are those that are prepared to actively intervene in economic processes to create conditions that provide the foundations for economic growth, in this case the 'subsidisation of industrialism' (Thompson 1996: 628) through import substitution industrialism (ISI).

ISI requires strong intervention from the government to create economic conditions that encourage local companies to manufacture goods for the domestic market rather than import them. Some of the interventions include placing tariffs on particular imports, manipulating exchange rates to favour local investment, subsidising local industry and awarding local companies industrial contracts over more established foreign ones. The ISI strategies adopted in Southeast Asia had qualified success, seeing the establishment of a number of consumer goods industries such as textiles, processed food, leather, pharmaceuticals and chemicals (see Dwyer 1990: 210). By manufacturing these higher value products locally Southeast Asian states lessened their reliance on core economies and were able to retain more wealth for further investment in country. However a problem with ISI strategies soon became apparent; the small size of domestic markets meant they were easily saturated, which limited the size and expansion opportunites of the new industries. As government protections only extended to their territorial space most fledgling industries could not compete with more established international competitors on global markets for quality or price. The ISI 'solution' to core–periphery dilemmas appeared to be a temporary and partial one at best.

A more lasting solution, for which the region would soon become famous, was to come from an unlikely source, the tiny city-state of Singapore. Singapore was originally a founding member of the Malaysian Federation that formed in 1963, but it was soon ejected when ethnic Malays began fearing rich Singaporean Chinese dominance within the politics and economy of the new state. As a result Singapore was left on its own with a very small domestic population, not enough to make ISI a viable strategy, but with a good educational system, an important port location and associated infrastructure, as well as strong post-colonial finance and trade links with Britain. Under the guidance of Prime Minister Lee Kwon Yew a new role for the developmental state evolved that saw the government create the conditions that would promote export-oriented industrialisation (EOI). EOI strategies involve the pursuit of FDI to create internationally competitive local industries. In Singapore this included investment in education to produce a skilled labour force, the suppression of labour unions to keep labour costs low, the setup of export processing zones (EPZs) in which normal labour laws, financial tariffs and regulations are relaxed, the provision of incentives such as tax holidays for foreign investors, and investment in transport, communications and financial infrastructure to ensure export efficiency for investors. The ISI strategy tried to protect domestic firms from foreign competition whereas EOI sought to attract foreign capital to build domestic economies. Singapore's EOI experiment surpassed all expectations as Western investors flocked to the country to invest in low cost manufacturing industries that would underpin Singapore's successful economic development.

Singapore's early economic success has been sustained, with temporary aberrations, through to the present time and has made the country an enigma in world development, being one of the few ex-colonial developing countries to have become as economically successful as core countries in Western Europe and North America. Only South Korea, Taiwan and Hong Kong, which with Singapore are known as the Asian tigers, can boast similar economic achievements. Their success challenged the logic of dependency theory as they showed that countries can trade their way out of the periphery, although whether they should be considered semi-periphery or bona fide core countries is a source of some debate. Similarly the claim that such economies represent positive examples of neo-liberal economic models should be treated with some caution given the important role of the developmental state in creating the conditions necessary to attract FDI; without such investment and market intervention foreign investment and growth is unlikely to have occurred (see Felker 2004).

Singapore's success attracted the attention of ASEAN-5 countries, which all soon adopted similar EOI policies to successfully attract FDI. While their conversion to EOI was not complete due to entrenched domestic interests protecting ISI strategies, it was not long before internationally competitive electronics, textile, food, steel and chemical industries began emerging in the industrialising cities of the region. North American and Western European institutions were the main initial investors, however these sources dried up during the oil shocks of the 1970s. FDI boomed again in the late 1980s when Japan and the Asian tiger economies became important additional investors, peaking in 1997 when US$33.8 billion or 7 per cent of global FDI was invested, a huge amount for a developing region (Pritchard 2006). The scale of the transformation can be seen in Malaysia and Thailand where 80 per cent of exports were primary commodities in 1980, however by 1995 Malaysia was exporting 80 per cent manufactured products and Thailand 74 per cent. In later chapters the enormous impacts and challenges associated with this industrialisation will be discussed, particularly the uneven impacts it has had on urban and rural spaces in terms of wealth, access to services and internal migration. These uneven impacts alongside the suppression of labour rights and political freedoms in two of the most successful EOI countries, Singapore and Malaysia, raise important questions about the equity of EOI strategies.

Development of state-led economies

State-led economies opted for a different development path based on socialist economic systems that were controlled by government interests rather than led by market demands. Such states often harboured strong suspicions about the motives of foreign actors and sought to protect themselves from neo-colonial exploitation. Burma, having watched British, Chinese and Indian communities profit from its resources under colonialism, opted for a unique isolationist development path entitled *The Burmese Road to Socialism* after a military coup in 1962. Northern Vietnam, having received no support from the West during its long anti-colonial war against France, also opted for a communist development strategy upon reaching independence in 1954, something it extended southwards during the US/Vietnam war (1964–1975). Laos adopted socialism when the Pathet Lao prevailed over the US-supported royalist government at the end of the Vietnam conflict and Cambodia began experimenting with state-led development, brutally under Pol Pot's regime from 1975–1978, and more conventionally under Vietnamese occupation from 1978–1989 (see Chapter 4).

The ultimate economic goal of socialist ideologies is to have the people, as opposed to private institutions, own and control economic production in an effort to discourage individualist profit-making and seek an equitable division of wealth. The state, as the representative of the people, can pursue this goal in different ways. In rural areas of Southeast Asia communist states claimed ownership of farmland and reallocated it to peasant households and communities. This aimed to overcome inequitable land ownership patterns, many of which had been set up during the colonial period and favoured 'friends' of colonial powers. Peasants were organised into collectives who would pool land, labour, equipment and resources in order to overcome the problematic economies of scale associated with small land parcels. Agricultural collectives would produce food for themselves and for the state with the proceeds of state sales divided among households within the collective according to the number of work hours they had committed. The state used agricultural produce to feed urban populations but would also support health and educational services within collectives, providing much greater rural access to these essential services than that which occurred within market-led economies.

Such rural reorganisation would seem to have much to do with the equitable development principles outlined in Chapter 1. Two of the guiding principles of socialist collectivisation include the equitable distribution of resources and the participation and empowerment of marginalised people (in this case peasant workers), however serious concerns can be raised regarding the curtailing of people's political freedoms and whether ultimately, collectivisation was leading to 'positive change'. Much depended on how the idealistic principles of socialism were put into practice. Vietnam underwent the most comprehensive form of collectivisation with the North almost entirely reorganised by the mid-1960s. Laos, in comparison, inherited a weak governing apparatus from the French and had little success in extending collectivisation beyond lowland central areas (Beresford 2001; Kerkvliet and Porter 1995). Rural reorganisation was often accompanied by violence with some 5,000–15,000 landowners and suspected traitors killed by overzealous Party officials in Vietnam and many thousands more sent to re-education camps (Brown, 1997: 248). Millions more tragically died in Cambodia when Pol Pot pursued an agrarian egalitarianism that forced people out of the cities to work agricultural land. However few consider Pol Pot's leadership, which was accompanied by thousands of executions and widespread starvation, to have anything to do with true socialist principles. A further problem was the low prices offered by the state for agricultural produce. These were far below international prices giving farmers little incentive to toil the fields for more than subsistence needs, limiting the amount of food available within the country. State attempts to stamp out private trade often resulted in disaster. The 1964 banning of non-state rice sales in Burma, for example, is blamed for an overall reduction in rice production and associated hunger a decade later (Brown 1997: 251).

Urban industries, like agriculture, were also controlled by the state in an effort to ensure the broader population, rather than private interests, shared in the benefits of industrial development. This was done by nationalising existing industries where the state took control and ownership of private companies, particularly foreign-owned companies that were formed by colonialists prior to independence. Factory owners were replaced by Party officials and Workers Committees that attempted to empower workers by giving them the chance to participate in the management and running of their workplaces (see Vu 2005). The state also invested in new enterprises. North Vietnam, for example, prioritised industrial development and built around 200 state-owned enterprises that produced things such as bicycles, water pumps, cast iron, fertilisers, cigarettes, antibiotics, cement and toothpaste, within its First Five Year Plan from 1961–1965. However the inefficiencies of rural areas were replicated in urban spaces with centralised decision-making structures slowing industrial production and state-owned enterprises having difficulty in accessing industrial inputs quickly and effectively. With socialism prioritising domestic self-sufficiency over foreign trade there were few export opportunities and few buyers given the strained relationships communist states had with wealthy Western nations. With foreign exchange hard to access it was difficult to buy foreign technology, such as machinery, making industrial development reliant upon aid from other communist nations such as the USSR and China. An additional hurdle for industrial development in Vietnam came during the American/Vietnam War when factories were specifically targeted by bombers and those that survived became reoriented towards supporting the war effort rather than economic development.

The inefficiencies within state-led economies soon saw them fall behind their market-led neighbours in terms of overall economic development. Communist states, despite comprehensive propaganda campaigns and recruitment programmes, struggled to retain the support of their populations. Many fled by sea as 'boat people' to seek refugee status in foreign countries while others sought refuge by crossing overland to Thailand. More passive forms of resistance took place on a daily basis with urban workers seeking out illegal small scale informal sector activities rather than devoting extra time to state projects for which they received little reward. In rural areas peasants prioritised their own fields for subsistence production over activities in farming collectives and sold surplus produce on black markets for higher returns than the state could offer. In the mid- to late 1980s these forms of internal dissent combined with a global decline in socialism led to the relaxation of state control of the economy and a greater openness to market-led initiatives. In 1986 Vietnam adopted reform policies known as *doi moi* (renovation) and Laos adopted the New Economic Mechanism (NEM). The newly named Myanmar (Burma) also attempted to make its economy more open after a change of personnel in the military government in 1989, yet international condemnation of its human rights record has slowed international investment and the extent of reforms taking place there. Socialism collapsed in Cambodia when Vietnam withdrew in 1989 and the country underwent economic restructuring under United Nations guidance.

Often referred to as *transitional economies* these countries are now experimenting with market-led approaches which, while still operating under state guidance, are much more closely aligned with the economic approaches of their Southeast Asian neighbours. Some of the reforms in Vietnam and Laos, for example, have concentrated upon attracting FDI through EOI, allowing markets to determine agricultural produce prices and encouraging privatisation. At the national scale the transitional period has undoubtedly been successful in stimulating economic growth. Vietnam, for example, has achieved economic growth rates of over 5 per cent since 1990 (Schaumburg-Muller 2005) and changed from a net importer of rice to be the world's third largest exporter in 1989 (Beresford 2001: 217). However at a local scale the unevenness that epitomises capitalist development would seem to be emerging in the socialist states that once prided themselves on principles of equity. In Vietnam, for example, *doi moi* has seen a decline of collectives that not only pooled labour and land but also provided cooperative health and welfare facilities, disadvantaging the poor and ill who were most reliant on those services (Beresford 2001). Similarly Rigg (1997) has argued that market reforms in Laos are benefiting wealthy, connected populations over poorer, remote ones. With Vietnam and Laos still professing socialist principles the challenge of ensuring the equitable distribution of the new economic benefits associated with transitional economies are considerable (see Box 3.1).

The Asian economic crisis: end of the miracle?

Southeast Asian economies were booming by the mid-1990s. In the words of Jonathan Rigg (2002: 137):

> In the years running up to 1997, a select band of East and Southeast Asian countries experienced perhaps the most rapid and sustained period of economic growth in human history. This growth was not a mere statistical sleight of hand: never had so many people been plucked out of poverty over such a short space of time.

This growth was unprecedented and much desired by development agencies as it showed that successful national scale economic development was possible, 'stockmarkets were booming, business was confident, and broad-based improvements in living standards were everywhere apparent' (Hill 1999: 1). With most other developing regions floundering since the end of colonialism East Asia became a source of interest for development theorists who sought out the secrets of its success. The World Bank's *East Asian Miracle* report claimed it was the region's adoption of free market principles and adherence to neo-liberal development philosophies that propelled its economies forward. In contrast neo-Marxists countered that the government had a strong role in intervening and improving domestic economies and it was this vital role of developmental states that was responsible for economic growth. Southeast Asian governments added a further dimension by stressing the unique cultural attributes of their populations, such as thrift and hard work, sometimes referred to as 'Asian values', that they claimed provided the region with unique

Box 3.1

Economic liberalisation, health services and Vietnam's doi moi

By the mid-1980s, the Vietnamese economy was in chaos. Inflation was at 400 per cent, chronic food shortages gripped the country and poverty was widespread. The country was on the brink of bankruptcy as a result of inefficiencies within the centrally planned economy, conflict with US, China and Cambodia, trade embargoes, and diminishing trade and aid support from the USSR. Disillusionment and frustration with state-directed economies led the ruling Communist Party to declare a set of economic reforms known as *doi moi* (renovation) that pursued market-oriented economies under socialist guidance. Vietnam since *doi moi* is one of the world's fastest growing economies, only surpassed by China.

While *doi moi* should be celebrated as a success it has been accompanied by new challenges and costs. Many of the negative components of market-led development in regional economies, such as growing inequality, urban primacy and a lack of formal sector jobs, are beginning to be felt here. Of particular concern are fears that the retraction in government spending will erase many of the unique achievements in education and health that were achieved under the state-planned economy. Prior to *doi moi* Vietnam had developed a comprehensive public health system that sought to provide free access to primary health centres in all agricultural communes across the countryside. This contributed to Vietnam having one of the highest life expectancies and lowest infant mortality rates for a country with its low level of economic development. Since *doi moi*, however, the retraction of government spending and break up of agricultural cooperatives have reduced funding from these traditional sources and Vietnam has seen the introduction of user-pays systems. For the wealthy, particularly those in urban areas where market forces are concentrating the best services, this has improved the quality of medical care that they can access but for the rural poor, the costs of medical care is often out of reach. A single visit can cost 8 per cent of the annual non-food consumption of a poor family and admission to a commune health station can cost 45 per cent of a poor family's annual consumption, pushing them into debt and poverty. A number of approaches such as compulsory health insurance and the waiving of fees for low income households has been discussed but the current impact of *doi moi* within Vietnamese health services is one of increasing inequality and concern.

Sources and further reading: Bloom (1998), Thoburn (2004)

advantages over its competitors (see Chapter 4). There was also a small group of post-development-type critiques that were less effusive about Southeast Asian progress, pointing out the inequitable nature of growth, 'environmental degradation on an unprecedented scale, growing inequalities between so-styled haves and have nots, rampant corruption, social malaise, and a widening in the gap between a prosperous core and a lagging periphery' (Rigg 2002: 137). Despite these disagreements there was a general consensus that many countries in East Asia, including most Southeast Asian economies (particularly the market-oriented economies but increasingly incorporating some of the transitional economies as well) had found a path out of underdevelopment.

The optimism that surrounded Southeast Asian economies came to an abrupt end in 1997 when the dangers of FDI-dependent growth became brutally apparent. What is now known as the Asian economic or financial crisis engulfed the region and spread to nearby countries with similarly structured economies such as South Korea and Taiwan. The crisis began in July 1997 when international currency speculators forced the floating and eventual collapse of the Thai baht (which was previously pegged to the US dollar), devaluing the economic assets of the country. International investors, in what has been described as a 'herd mentality' left the country in droves, or in a 'panicked scramble to withdraw capital before the inevitable devaluation' (Felker 2004: 68). With the Thai economy in freefall investors feared spill-over effects into the wider region and began withdrawing their funds from other market-led Southeast Asian economies and investing in safer locations such as China. The region was thrown into crisis as the value of local currencies dropped, foreign investment dried up and GDP growth slowed or reversed (see Figure 3.2). Just as the Southeast Asia had stunned the world by its rapid growth during the miracle years, it stunned the world again with its spectacular collapse during the economic crisis.

The impacts of the crisis were felt unevenly across Southeast Asia. The market-oriented economies were the worst hit, as they were more reliant on FDI, with Indonesia and Thailand being the most severely affected. In 1998 Indonesia's economy fell as much in one year as that which occurred over three years at the peak of the Great Depression in the United Kingdom, its currency at one point dropping to one-sixth of its pre-crisis levels (Hill 1999). With the value of the currency plummeting (it dropped 25 per cent in one day) overseas debts, when calculated in the local currency, became impossible to pay, causing

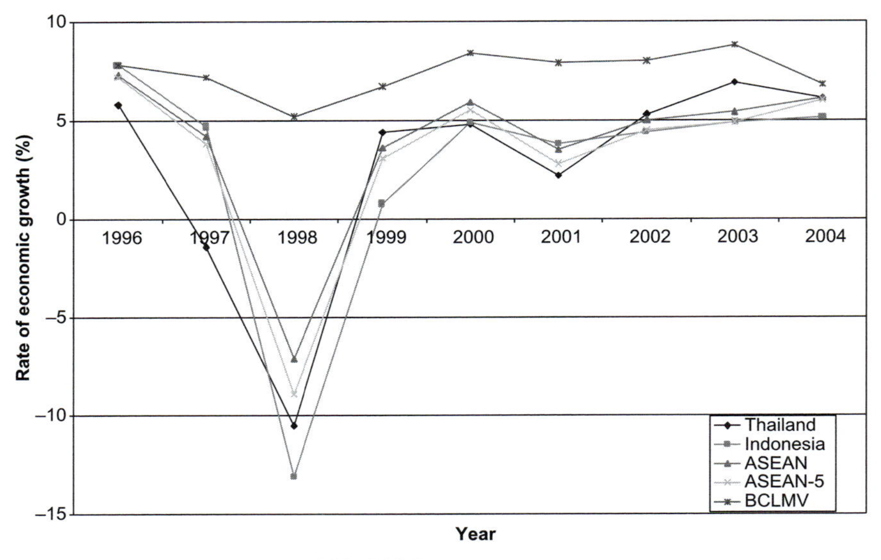

Figure 3.2 ASEAN growth rates 1996–2004.

Source: ASEAN Secretariat (2005a)
Note: BCLMV refers to the combined economies of Brunei, Cambodia, Laos, Myanmar, Vietnam.

defaults in loan repayments and the further departure of foreign investors. Key industries collapsed and financial institutions went out of business leading to a surge of unemployment, particularly among white collar formal sector workers but also among those on the factory floor (see Beaverstock and Doel 2001). Poverty rates in Indonesia are estimated to have risen from 10.1 per cent before the crisis to 14.1 per cent by World Bank estimates, to 39.9 per cent by Indonesian Central Board of Statistics estimates, or to 66 per cent by International Labour Organisation estimates (Booth 1999). If the middle Indonesian estimate is correct an additional 57 million people were propelled into poverty during this period.

The decision of international investors to pull out of Southeast Asia was to have very real effects on the lives of ordinary people. The urban middle classes were the hardest hit as they were the most dependent upon global economic flows. Bankers, managers, accountants, clerks and technicians suddenly found themselves without a job as did thousands of blue collar workers whose factories or building sites could no longer afford to import industrial inputs. This had downstream ramifications for sales, hospitality and a range of other services whose workers were suddenly faced with a bleak future. The misfortune experienced by the unemployed was sharpened by the corresponding rise in inflation that devalued any savings they may have had while increasing the price of basic food products. In Indonesia basic food prices increased 133 per cent on average but are reported to have quadrupled in some places causing hunger and hardship (Booth 1999; Silvey 2001). In the space of just a few weeks some people went from secure middle-class lifestyles to unemployed job-seekers with few prospects, little money and growing difficulties in even feeding their families. There were little government funds available to assist the legions of newly unemployment however work-for-food welfare programmes provided a base level of support for some in Indonesia. Instead many workers found financial 'safety nets' within informal economies where they became absorbed as small scale family enterprises, such as the production of street-stall food or returned to rural areas where they worked on family farms (see Box 3.2). Many however fell through the cracks and desperation led to anger and disenchantment. In Indonesia riots wrecked urban areas, particularly Chinese businesses, and eventually contributed to the resignation of President Suharto.

Box 3.2

Mr Sandwich: foodscapes and safety nets during the Asian economic crisis

Perhaps the most famous victim of the Asian economic crisis is the former high profile Thai stockbroker Sirivat Voravetvuthikun who is now 'hundreds of millions' in debt because of a failed investment in a 28-unit condominium project. Sirivat has survived the crisis by finding work in the informal small scale food vending industry, in his case selling sandwiches. With his slogan 'I'd rather be bankrupt than dead' he has come to personify Thai ingenuity and resilience in the face of the economic meltdown. He is now known as Mr Sandwich, has his own website and offers motivational speeches to others who have been affected by the economic downturn. Sirivat's case has been mirrored by many less famous victims of the crisis with the informal food sector requiring few skills or costs to start up and providing a quick attractive low cost option for urban consumers. With the devaluation of local currencies the costs of middle-class restaurants and Western fast food chains became exorbitant; at the height of the crisis a burger cost 10,400 rupiah (US$1.20), the equivalent of two days' work for most Indonesian workers. In contrast a street dish made of rice, meat and eggs could be purchased for 2,000 Rp or less.

Small scale food vending has also provided an important income source for migrants who, having lost their jobs, have been forced to return to their rural roots:

> Life is hard for 34-year-old *Mbak* (sister) Wok since she lost her beloved husband Poniman . . . and her job [in Surabaya] three months ago. Employed for two weeks in an irrigation project in her village Sumur Welut, she earns a daily wage of Rp. 6,500 [US$0.70] to feed her five children. 'I have taken any jobs I could get my hands on after my husband died in January [1997] because I have to raise my children alone. I have nobody else to help us'. Her children have been forced out of school because her income is hardly enough to meet their daily needs, let alone pay for school fees . . . She also sells *bakso* (meatball soup) at night for additional income.
>
> (Jakarta Post, 4 October 1998, cited in McGee and Firman 2000)

Small scale food vending operations have provided an unlikely safety net for victims of the economic crisis and temporarily rearranged foodscapes in the region. Their vital role in sustaining livelihoods and providing food security during the crisis suggests local authorities should encourage such enterprises and assist them with health and safety training and education rather than threaten them, as is more common because of their unlicensed and unregulated nature, with eviction and demolition. Selling food on busy streets can be a polluting and hazardous occupation; the more support people like Mr Sandwich can receive the safer it is for him and the broader economy.

Sources and further reading: Yasmeen (2001), Bacani (1997), McGee and Firman (2000)

Rural and remote areas of market-led economies were generally less affected by the crisis than urban areas as rural economies are less dependent upon international financial markets for FDI. In some cases the collapse of domestic currencies made export crops more competitive, as the international market could buy them for cheaper prices as long as imported input costs, such as international pesticides and fertilisers, were minimised (see Booth 1999). This has led some commentators to suggest that one of the ironic impacts has been to lessen inequality throughout the region as wealthier countries, urban centres and households experienced greater absolute financial losses over poorer ones (see Akita and Alishjahbana 2002). However this hardly equates to equitable development as it is pulls economically successful areas back rather than positively propelling poorer ones forwards. Rural spaces did undergo many changes, mainly as a result of back migration of urban workers to their rural roots and because of the spiralling costs of basic goods. The worst affected were the poorest people within society who struggled to cope with increased prices. In some cases poor households were forced to withdraw their children from school because of the costs of textbooks and uniforms and employed them on family farms to increase agricultural output.

Women were also unduly affected as Silvey (2001) has documented in her study of young women returning to their gendered Indonesian rural homes in South Sulawesi, Indonesia. Young female returnees were expected to queue and wash clothes at public water pumps for longer periods due to the rising cost of soap; spend longer hours at markets bartering due a general reduction in rural incomes; seek out less accessible firewood due to rising kerosene prices to cook cheap family meals; and inherited the responsibility of caring for sick families members who were more frequently ill due to hunger and the increasing costs of visiting health clinics. In contrast young returnee men were expected to wait out the crisis before returning to formal sector work. In the words of one disgruntled female returnee, 'He just sits around because he's unemployed. We all have to help out because it's a monetary crisis, but for me [the workload] is double. Also, he keeps smoking cigarettes . . . and that costs money we don't have' (Silvey 2001: 41–42).

The economic crisis threatened to reverse the developmental gains of the world's most successful developing region, throwing doubt on claims that developing countries could escape their peripheral status and grow into successful independent states. Rigg (2002) has observed how quickly neo-liberals, neo-Marxists and culturalists that had earlier claimed responsibility for the region's pre-crisis growth modified their arguments to blame alternative approaches for the weaknesses of pre-crisis economies. Hence the IMF saw no contradiction in offering to lend money to struggling countries in return for further neo-liberal reforms, despite many other commentators blaming free market activities in the form of FDI-dependence for the crisis itself. The IMF blamed corruption and 'crony capitalism' for the economic crisis, believing Southeast Asian governments were too interventionist in markets by offering guarantees to government-linked businesses and protecting special interest groups, thereby reducing market efficiencies. Their solution was to enforce structural adjustment packages (SAPs), or what are now known as poverty reduction strategy papers, which are neo-liberal programmes oriented at increasing economic efficiencies and making the region more attractive to international investors. Both Indonesia and Thailand agreed to SAPs in a desperate effort to stabilise their economies while the Philippines received an extension of an earlier agreement.

Indonesia is generally considered to have undergone the most extensive and inflexible SAP with the IMF only releasing funds upon receipt of monthly reports detailing which of the IMF's schedule of policy reforms had been implemented. Such tight monitoring effectively put the IMF in command of the Indonesian economy raising important concerns about questions of sovereignty and autonomy within countries undergoing IMF restructuring. The local impacts of the reforms were severe: a reduction in government spending led to the retrenchment of thousands of government employees and a retraction of public services; privatisation raised the price of many services putting them out of reach of the poor; increasing interest rates put loans out of reach of small businesses; and a reduction of import, export and ownership restrictions encouraged the entry of foreign competitors into the domestic marketplace. This increased the likelihood of struggling local businesses being purchased by foreign capital for bargain prices, and added to environmental degradation as government regulations and taxes were relaxed (see Grenville 2004). Fears of these types of impacts saw Malaysia decline the IMF's offer of support while Singapore's economy was strong enough to rebound quickly and lead the recovery of the region.

By 2002 the worst of the crisis had passed and growth returned to pre-crisis levels. This has led some to question whether the crisis was a temporary aberration in the overall economic development of the region or whether it reflected some more fundamental weaknesses in regional economies. Either way the crisis has left its imprint upon the economies of the region and exposed weaknesses in the resilience and sustainability of FDI-led industrialisation strategies.

Post-crisis economies

The post-crisis economies of Southeast Asia are marked by diversity and unevenness. Low income economies remain reliant on natural resources and the agricultural sector. Timor-Leste, for example, is currently negotiating with Australia over access to massive oil reserves in the Timor Sea to boost its undeveloped economy. Oil income is being put into a petroleum fund to diversify its economy but it remains to be seen whether Timor will be able to develop internationally competitive industries or whether it will become overly dependent on a single income source as is the case in Brunei (see Chapter 8). Laos has attracted controversy for its ambitious plans to invest in large hydropower schemes within the Mekong River basin, which will flood large parcels of lands and force thousands from their homes, and sell the electricity to Thailand. Western investment in Myanmar is unlikely to improve until improbable political reform takes place leaving its industrial future dependent upon attracting investment from China and India, both of whom appear less concerned about the government's human rights record. Poverty is common in Cambodia, which is struggling to find its niche but hopes to emulate the successes of its former occupier and neighbour, Vietnam, which has experienced rapid economic growth since *doi moi*. By opening its economy to market reforms and offering a cheap well-controlled labour force Vietnam successfully attracted FDI and now has one of the world's fastest growing economies.

Of the ASEAN-5 nations the post-crisis recovery has been slowest in Indonesia where the economic downturn and accompanying political crisis weakened state control of the economy and created uncertainty in the minds of overseas investors. Growth is positive but ethnic violence and a series of disasters including the 2004 Asian tsunami, the 2005 Nias earthquake and two further earthquakes in Java in 2006 have slowed recovery. The Philippines has one of the slowest economies and has resorted to exporting people overseas to earn foreign incomes, a proportion of which is sent home to their families. A significant proportion of its GDP is now coming in the form of remittances from Filipino nurses, maids, nannies and construction workers who are employed overseas (see Chapter 5). Thailand experienced rapid uneven growth under the free market principles of former Prime Minister Thaksin; however he has was removed during a bloodless military coup in 2006. Its important tourism industry, the largest in Southeast Asia, has been shaken by the coup and Buddhist/Muslim conflict in the south of the country, as well as bad regional publicity related to SARS, bird flu and the devastating Asian tsunami of 2004. Malaysia has sought to distinguish itself from other Southeast Asian economies by investing in the 'Multimedia Super Corridor' which boasts world-class digital communications infrastructure in the hope it will one day mimic the high-technology innovations and knowledges normally associated with the US's Silicon Valley (see Bunnell 2004). At the same time it is cutting down large swathes of forest as it plants new oil palm plantations, which many believe will be a key bio-fuel of the future. Brunei is almost entirely dependent upon the sale of oil, petroleum and natural gas products to sustain its economy while Singapore remains the wealthiest economy in the region, investing in education and transport infrastructure to ensure it retains its position as the core financial and service centre hub of the region. In addition it is attempting to make the city more appealing to expatriates and offers low tax rates for highly skilled workers in an attempt to build high quality knowledge industries.

Pritchard (2006) has argued that the post-crisis FDI-based economic development strategies of Southeast Asia are considerably more risky than they were in the past. Some of the key conditions have changed including a global downturn in FDI, the emergence of free trade zones such as the European Union (EU) and the North American Free Trade Agreement which keeps FDI closer to home, and the emergence of significant low cost FDI-competitors, most notably India and China. India's education system and English colonial heritage is thought to be partly responsible for its success in securing new FDI-industries such as call centres that could otherwise have gone to Southeast Asia while China's low cost labour, huge domestic market and booming economy has made it attractive to overseas investors. In 2003 China attracted US$53 billion in FDI, much of which could otherwise have been invested in Southeast Asia, which only received US$19 billion (Frost 2004; Pritchard 2006). The fear of competition is expressed by Prime Minister Goh Chok of Singapore in 2001: 'Our biggest challenge is to secure a niche for ourselves as China swamps the world with her high-quality but cheaper products . . . How does Singapore compete against 10 post-war Japans, all industrialising and exporting at the same time' (Asiaweek, 31 August 2001, cited in Felker 2003: 257). While China may be casting a shadow over the long-term economic sustainability of FDI-led development it may also provide the source of solutions as it grows in economic power and influence. ASEAN–China trade already tops US$78 billion and this will increase if proposed free trade agreements are realised in the future. In addition ASEAN countries attract a greater share of Chinese investment than the US, EU, Hong Kong or Macau. Singapore, for example attracted US$640 million worth of investment in 1990–1999 while in the same period Thailand had 498 Chinese firms working in country (Asiaweek, 31 August 2001, cited in Felker 2003: 257). Clearly China will play a key role in the future of Southeast Asian economic development, either as a competitor for FDI or an FDI investor, and probably both.

A defining characteristic of post-crisis economies is the uneven distribution of wealth and associated access to services at regional, national and local scales. The average income in Laos, for example, is around 60 times less than that of Singapore, 11 times less than Malaysia and 6 times less than neighbouring Thailand. Similar differences can be seen within country borders where successful urban middle classes capture the majority of the economic benefits of development over their more remote rural cousins. While economic wealth is only one component of equitable development it does impinge on many other attributes, such as access to health and education, livelihood opportunities and social empowerment. The life expectancies of those in Laos and Timor-Leste, for example, at 55 years is some 20 years less than those living in the wealthier countries of Singapore and Brunei, with almost 10 per cent of children dying before the age of 1 (see Table 1.1). These types of figures are replicated at smaller scales and signify the importance of overcoming uneven distributions of wealth. The remainder of this chapter will consider these challenges at regional and national scales.

Regional inequality

There are no easy solutions for overcoming the regional unevenness that characterises economic development in Southeast Asia, however various initiatives are being explored. One is to promote neo-liberal approaches through the ASEAN Free Trade Agreement (AFTA). AFTA was originally formed by the market-led economies of Singapore, Malaysia, Thailand, Indonesia, Brunei and the Philippines in 1992 with Vietnam joining in 1995, Laos and Myanmar in 1997 and Cambodia in 1999 (as they joined ASEAN – see Chapter 4). AFTA has dual aims; the first is to attract more FDI to the region by minimising the costs of investing in several countries at once. Hence a transnational corporation (TNC) could have the rubber soles of shoes built in one country, the leather top in another and the shoe glued in a third country without paying tariffs when shifting products between countries. Second, AFTA seeks to increase intra-regional investment and trade. Colonial trading systems, the division between state and market-led economic economies, and EOI-development strategies have resulted in Southeast Asian countries being much more outward looking and reliant upon external trading partners than the economies of their neighbours. In breaking down tariffs and guaranteeing regional investors the same rights as domestic investors AFTA aims to facilitate further investment from high income countries such as Singapore in the low income economies of the region (see ASEAN Secretariat 2003; Nesadurai 2003).

While AFTA has facilitated free trade in various sectors there are many industries that are still protected for fear that they will become unprofitable when faced with open regional competition. This has slowed the establishment of a regional free trade area and some countries, such as Thailand and Singapore, have put their efforts into bilateral free trade agreements with external countries, slowing regional progress. Within the region some innovative smaller scale inter-country collaborations have been attempted that seek to build on the competitive advantageous of each partner country. The most famous of these is the Indonesia–Malaysia–Singapore Growth Triangle that draws cheap labour and land from Indonesia, skilled labour and land from Malaysia, and finance and technical expertise from Singapore (see Box 3.3). The long-term effects of these types of initiatives is unknown and much depends on whether neo-liberal free

Box 3.3

Borderless worlds? The Indonesia–Malaysia–Singapore Growth Triangle

The Indonesia–Malaysia–Singapore Growth Triangle (IMS-GT) was a concept announced on 20 December 1989 by then deputy prime minister of Singapore, Mr Goh Chok Tong, and formally signed off by the three nations in 1994. The guiding principle behind the plan was to link three geographical areas of Singapore, Malaysia and Indonesia into a larger economically integrated regional unit. Each country brings something unique to the triangle: the Malaysia state of Johor, which forms the head of the triangle, contributes land, semi-skilled labour and good infrastructure; Singapore in the middle contributes capital, technology, finance and management skills; while the Riau Islands of Indonesia form the base contributing unskilled low cost labour, land and access to Indonesian oil and gas reserves (see Figure 3.3). The IMS-GT allows for the easy movement of capital, goods and services within the borders of the triangle and has been lauded as a step in the direction of a 'borderless' world.

Economic growth within the IMS-GT is having particularly significant impacts on the Indonesian islands of Batam and, to a lesser extent, Bintan. There is now a range of industrial parks situated on the islands that have attracted high levels of investment from Singapore and elsewhere. The economic attraction of the Riau Islands is described approvingly by *The Economist* magazine as follows:

> The Batam Industrial Corporation (Batamindo), 40% owned by Singaporean interests and 60% by the Salim Group is transforming the island of Batam into a floating factory, two-thirds the size of Singapore and only a 30-minute boat trip from its financial district . . . Batamindo will supply custom-built factories on 30-year leases . . . The Salim group can supply unskilled but nimble-fingered workers (mostly young girls) at S$92.50 ($54) a month.
>
> (*The Economist* 16 November 1991: 9, cited in Sparke *et al.* 2004)

Figure 3.3 Indonesia–Malaysia–Singapore Growth Triangle.

The low labour costs have attracted capital that has transformed what was a once a quiet rural fishing economy into an industrial enclave that attracts migrants from across Indonesia.

The claims of a borderless world, however, are limited to unidirectional flows of Singaporeans and Singaporean capital into Indonesia, but few opportunities for reverse labour flows. The devaluing of the Indonesian rupiah during the Asian economic crisis as well as tight customs and border policing restrict Indonesian movements into Singapore. As Rizal, a Bintan islander laments:

> I'd really like to go to Singapore to see my mother's family, but I'd need a lot of money to make the trip because our money isn't worth anything. If I took my money to Singapore, I couldn't do anything. But I'd still like to go there.
>
> (Ford and Lyons 2006: 263)

In contrast Abdullah, also from Bintan, says:

> The crisis was no problem for Singaporeans. They are better off because the exchange rate is better for them. Even if they come . . . with only a few dollars, they can go shopping, stay in a flash hotel, and enjoy luxurious facilities.
>
> (Ford and Lyons 2006: 263)

Ironically it is the same economic conditions that have attracted the borderless flows of capital from Singapore, i.e. low wages, that prevent the borderless flow of labour back into Singapore. The IMS-GT has succeeded in sparking regional economic growth but does so by exploiting the economic advantages *and* the economic disparities of the region.

Sources and further reading: Bunnell *et al.* (2006), Sparke *et al.* (2004), Ford and Lyons (2006), Grundy-Warr *et al.* (1999)

markets will boost low income economies. If, however, dependency theories are more accurate, the establishment of free trade initiatives may simply entrench the uneven nature of Southeast Asian economic development with a new intra-regional core–periphery model evolving in which richer countries exploit the resources of the poor. Thailand's decision to protect its own forests, for example, has led to an increase in deforestation in neighbouring countries as it sources wood products from there. Similarly free trade may lead to worse rather than improved wages and working conditions. If any of the ASEAN-5 nations wish to compete with Vietnam in attracting intra-regional FDI they will be tempted to suppress wages and labour conditions to become more competitive, effecting creating a competitive 'race-to-the-bottom' rather than positive change.

National scale equality and inequality

National scale economic inequality is just as pronounced as intra-regional inequality. The most common way of measuring this is through the Gini Index, which shows how effectively income is distributed within a country. A score of 0 equates to perfect equality with everyone earning the same income, while a score of 100 indicates perfect inequality – one person earns all income. Table 3.2 shows that

Table 3.2 ***Inequality measures (multiple years)***

Country	*Gini Index*	*Share of national income of poorest 20%*	*Share of national income of richest 20%*	*Average income ratio of poorest 20% to richest 20%*	*Population living below US$2 a day (%)*
Singapore	42.5	5	49	9.7	n/a
Brunei	n/a	n/a	n/a	n/a	n/a
Malaysia	49.2	4.4	54.3	12.4	9.3
Thailand	42	6.1	50.0	8.3	32.5
Philippines	46.1	5.4	52.3	9.7	46.4
Vietnam	37	7.5	45.4	6.0	n/a
Indonesia	34.3	8.4	43.3	5.2	52.4
Myanmar	n/a	n/a	n/a	n/a	n/a
Cambodia	40.4	6.9	47.6	6.9	77.7
Laos	34.6	7.6	45.0	6.0	73.2
Timor-Leste	n/a	n/a	n/a	n/a	n/a

Sources: UNDP (2005, 2006)

Note: n/a – no statistics available.

the state-led development paths of Vietnam, Laos and Cambodia have, with the exception of the sprawling archipelago of Indonesia, been more effective in their goals of pursuing income equality than their market-led neighbours. However a high percentage of people in these more equitable economies live on less than US$2 a day, which suggests that state-led development has resulted in ' "shared poverty" rather than rising overall living standards' (Beresford 2001: 211–212). The richer countries appear to have traded income equality for overall increases in wealth, although the Philippines fares badly in all categories, having almost half its population living on less than US$2 a day and a high Gini score. The roots of the Philippines misfortune derive from the inequitable land ownership patterns that were established during colonial times, something it is trying to overcome through land reform programmes, as will be analysed in Chapter 7. National income inequality also reflects the evolution of class structures within the market-led economies (see Chapter 5) and the concentration of higher paying service and manufacturing industries in urban centres. This spatial inequality has caused mass migrations of people from rural to urban areas in search of higher paying employment, changing the fabric or urban and rural societies as will be addressed in Chapters 6 and 7.

One of the challenges that emerges from the data in Table 3.2 is whether the countries currently undergoing transition can maintain their more equitable distribution of income as they open up to more market-led approaches. Or alternatively, is income inequality an intrinsic feature of market-based economies and, if it is, does it matter as long as everyone is getting wealthier, even if some are increasing their wealth faster than others? The answer to the latter question depends on your personal philosophy but current indications suggest that markets are likely to increase rather than decrease income inequalities in transitional economies. Rigg (1997), for example, argues that the benefits of greater market liberalisation are likely to be concentrated in lowland areas (predominantly from the Lao Loum ethnic group) and those more entrepreneurial classes, many of which are ethnic Chinese, Vietnamese or Thai. The benefits, if any, of increased market access for remote highland Lao people (predominantly ethnic Lao Soung who are traditionally the poorest population) are likely to be minimal – thereby increasing, rather than decreasing, national inequality. Hughes (2003: 32–33) reports a similar process in Cambodia where 'the move to a free market significantly increased social stratification, enriching those in power, particularly those with power over privatisation of land and resources, and created large groups of marginalised and property-less poor'. If this is the case the developmental state still has a strong role to play in intervening in market activities in attempts to secure more equitable forms of development.

Conclusions

The diversity of Southeast Asian economic landscapes reflects the economic conditions new states inherited upon independence and the subsequent post-colonial development paths they adopted. It has led to extraordinary diversity in economies and great unevenness in the distribution of wealth and services. Market-oriented economies have fared the best in terms of overall GDP growth and poverty reduction, but possess large disparities in income. State-led approaches have struggled to match the GDP-growth of the market-led economies but are becoming wealthier, and possibly more inequitable, as they increasingly engage with international market systems. The economic development path countries in Southeast Asia appear to be converging upon, one in which developmental states play a strong role in nurturing the economy through FDI-oriented strategies, has been successful in building economies but it comes with risks as evident in the Asian economic crisis. The future of this form of economic development is further clouded by the emergence of China as a major FDI competitor. At the national scale many countries of Southeast Asia have benefited from economic development, however there are many challenges ahead, particularly in terms of sharing the benefits of that wealth more evenly throughout regional and national economies.

Summary

- **Southeast Asian states have pursued a variety of economic strategies in the independence period, all of which have been guided, to different extents, by 'developmental states'.**
- **States that pursued market-based growth strategies based on EOI have outperformed other states in terms of overall GDP.**
- **States that have pursued state-led economies have struggled to grow their economies but have done so in more equitable ways.**
- **The Asian economic crisis threw the economies of the region into turmoil and resulted in some countries undertaking neo-liberal reforms in order to access loans from the IMF.**
- **The current economic landscapes of Southeast Asia are marked by unevenness at all scales, posing considerable challenges for equitable development.**

Discussion questions

1. Compare and contrast the economic development of state- and market-led economies from the perspective of equitable development.
2. Explain the importance of FDI to the region.
3. Consider whether Southeast Asia is likely to become more or less equitable, in terms of the distribution of wealth and services, in the future.

Further reading

For a discussion on how Southeast Asia's economic development was theorised before, during and after the Asian financial crisis see:

Rigg, J. (2002) Of miracles and crises: (re-)interpretations of growth and decline in East and Southeast Asia. *Asia Pacific Viewpoint* 43(2): 137–156.

For an assessment on the future of FDI in Southeast Asia see:

Pritchard, B. (2006) More than a 'blip': the changed character of South-East Asia's engagement with the global economy in the post-1997 period. *Asia-Pacific Viewpoint* 47(3): 311–326.

For a detailed history of the different development paths Southeast Asian countries pursued after during and colonialism see:

Brown, I. (1997) *Economic change in Southeast Asia c. 1830–1980*. Kuala Lumpur: Oxford University Press.

Useful websites

The official ASEAN website that has up-to-date information on happenings in the region and annually updated economic data: www.aseansec.org/

Further economic information can be sourced from the World Bank: www.worldbank.org

4 Political development

Introduction

Global political development is becoming increasingly influenced by ideas of 'good governance'. Multilateral development institutions, international aid donors, domestic political parties and local constituencies within and without Southeast Asia all express support for the concept. However despite, or perhaps because of, the popularity of good governance its actual meanings are elusive and seem to shift according to place and context. The term originated from Africa in the late 1980s where scholars called for better state–society relations based upon development, democracy and socially inclusiveness or participation. Since then the term has been used in many different ways, most significantly by neo-liberal theorists who often blame 'poor governance' for the lack of economic development in a country after it has undergone neo-liberal economic reform. For these theorists good governance refers to accountability and transparency within government processes, thereby securing more favourable investment conditions for foreign companies (see Mkandawire 2007). Post-development theorists are sceptical of such claims and are more likely to see the concept of good governance as a new discursive tool by which the development industry is attempting to manipulate and reshape developing countries. Good governance reforms were imposed on Indonesia during the economic crisis, for example, which forced the country to rearrange its governance structures in ways dictated by IMF, rather than local, guidelines.

From the equitable development perspective adopted in this book good governance is associated with increasing political freedoms and empowering people to participate in governance systems. While such goals are appealing to most the means of pursuing them are inherently controversial. Some people believe, for example, that democracies are the best means of increasing political freedoms and promoting participation and push for democratic reforms in the name of political development. Others may counter that democracies tend to legitimise majority rule over minorities and instead favour decentralisation and local political autonomy as more effective forms of good governance. Communist states may promote the belief that the only truly equitable form of governance is a socialist state in which the people are, at least in theory, the state. Even repressive authoritarian states enter good governance debates by claiming they are guardians of their people and are the most efficient and appropriate rulers to govern the state in the people's interests. This chapter explores the different approaches Southeast Asian states have adopted in their pursuit of political development and the variety of 'good governance' systems that have evolved.

International influences on political development

As discussed in Chapter 2 many of the foundations of contemporary Southeast Asian states were established during the colonial period when Western empires introduced modern systems of governance. It was during this time that many state apparatuses evolved, such as police forces, judicial systems, tax regimes, education systems, medical infrastructures and political bureaucracies, and were inherited by post-colonial governments. While many of these processes were 'indigenised' upon independence, post-colonial states had far more in common with their colonial forerunners than they did with earlier indigenous pre-colonial kingdoms. The varying strength of these governance systems reflected the amount of resources colonial powers invested in governance during the colonial period and the extent of damage they incurred during the Second World War and subsequent independence struggles. It is generally agreed, for example, that France invested far less resources in the governance of Laos and Cambodia than they did in Vietnam, resulting in weak states in these former French territories. Similarly the nine-year war of independence Vietnam was forced to wage against its French occupiers (1945–1954) resulted in the destruction of many colonial governance systems that had to be redesigned and rebuilt in the years that followed. Indonesia underwent similar rebuilding after fighting Holland for independence from 1945–1949. However the worst war-affected country has been Timor-Leste, which declared independence from Portugal on 28 November 1975 only to be invaded and colonised by Indonesia on 7 December of the same year. After a long and violent guerrilla war Timor-Leste secured its independence in 1999, but not before Indonesian militia destroyed much of the physical and political infrastructure required for governing. So thorough was the destruction that the United Nations effectively governed the country as structures were rebuilt and full independence announced in 2002. In contrast the peaceful handover of power by the US to the Philippines and from Britain to Burma and the Malaysian Federation saw indigenous elites inherit colonial governance systems intact, benefiting their political development (see Figure 2.3).

One of the most important legacies of the colonial period has been the establishment of what have proved to be extremely resilient territorial borders. With the exception of Indonesia's occupation of Timor-Leste (1975–1999) and Vietnam's occupation of Cambodia (1979–1989), both of which have now ended, the political boundaries drawn by colonial powers remain largely intact today. There have been no significant international wars between countries in the region since independence, although Indonesia came close when it adopted a policy of *Konfrontasi* with Malaysia from 1963–1966. *Konfrontasi* was an attempt to prevent the formation of the Malaysia Federation via 'a practice of coercive diplomacy, employing military measures stopping short of all-out war, which was designed to create a sense of international crisis in order to provoke diplomatic intervention in Indonesia's interest' (Leifer 1995: 80). During *Konfrontasi* Indonesia applied diplomatic pressure, mobilised troops along Malaysia's borders and threatened to 'crush' Malaysia, which it saw as a neo-colonial creation of Britain, if it didn't conform to an alternative regional confederation that brought the nations of Malaysia, the Philippines and Indonesia together under the umbrella structure of Malphindo (see Woodard 1998). *Konfrontasi* failed to achieve its aims and the territorial borders of these countries have remained intact ever since (see Figure 4.1).

The second major external influence upon political development in Southeast Asia was the ideological Cold War battle between the liberal democratic principles and capitalist systems promoted by the US, and the communist systems and socialist values espoused by the USSR.

Figure 4.1 Indonesia, Malaysia, Brunei and Mindanao (the Philippines).

Both were battling for global political hegemony and with Southeast Asia positioned south of the world's largest communist state, China, and east of the world's largest democracy, India, it was a region of considerable strategic importance. Leaders within Southeast Asia were forced to align themselves with one of the superpowers if they sought support in the form of trade and aid. The ASEAN-5 states aligned themselves with the US while North Vietnam, which did not trust the West after its long war of independence, aligned itself with the USSR. This troubled the US who used a falling domino analogy to explain their fear that if Vietnam 'fell' to communism the ideology would spread to the rest of the region. To prevent this from occurring the US engaged in war to prop up a non-communist (and locally unpopular) regime in the south of the country but failed to win a war that stretched from 1965–1975. Millions of civilians and hundreds of thousands of soldiers died during the conflict that stunted any attempt at economic or political development.

Over the border in Laos a 'secret war' was taking place in which the Royal Lao Government Army and a secret army of CIA-funded Hmong guerrillas were engaged in a battle with the communist Pathet Lao, which was supported by the USSR and Vietnam. The Pathet Lao would eventually emerge victorious after 1975 although it came with huge costs. Conflict forced thousands of refugees from their homes and US-bombing, which resulted in it becoming the most bombed country per capita in history, has lingering impacts on development today (see Box 4.1). Further south the Vietnamese helped

Box 4.1

The Secret War, UXO and Lao development

While the war in Vietnam officially ended in 1975 the impacts upon development are still being felt throughout the former Indochinese nations. Laos, for example, never officially featured in US war documents, instead it was euphemistically referred to as 'the other theatre', yet it attracted sustained US bombing and is now the most bombed country per capita in the history of the world. A secret war took place in Laos between the US-supported Royalist Government and the communist Pathet Lao located in the rural northeast of the country. Internal peace talks about a coalition government broke down when the US threatened to withdraw their aid funding, which supported the salaries of the entire Royal Lao Army as well as a secret CIA-funded ethnic Hmong army, if communist talks went ahead. Instead the US funded an unsuccessful air and ground campaign that cost thousands of lives and forced hundreds of thousands of people from their homes, villages and fields.

One of the most significant lingering impacts of the war are the small tennis-ball-sized 'bombies', as they are known in Laos, that remain embedded in the landscape. At least 30 per cent of these bombies did not explode on impact and continue to cause up to 300 accidents each year, to mostly farmers and children, who are seriously injured or killed. The bombies are slowing development in Laos as farmers are afraid to clear new fields and new initiatives such as roads, forestry and dams must wait for over-worked bomb clearance teams to give the go ahead (see Figure 4.2). While aid money is flowing into the country to fund UXO Lao, the official government organisation responsible for managing UXO (unexploded ordinance) risks and clearance activities, as well as international NGOs such as Handicap International and the Mine Action Group, the degree of contamination is so heavy, and the funding so limited, that it is unlikely that Laos will ever be fully free of bombies. There is a global campaign to ban cluster bombs for these long-term ongoing destructive impacts.

Figure 4.2 ***UXO Lao information poster.***

Source: Author

Meanwhile widespread poverty in Laos has encouraged people to take advantage of the situation. Spent bomb casings are incorporated into building designs or used as fences or tables in rural areas. A more recent worrying trend has seen people attempt to decommission live bombs in order to sell the scrap metal to dealers. Rumours suggest Vietnamese traders are providing children with metal detectors to encourage them to carry out this task. Such high risk activities are of immense concern to authorities and the aid community who have responded by illegalising the scrap metal trade, and introducing UXO education programmes to the population through puppet shows, radio spots and the school curriculum. However, the extreme poverty in some parts of rural Laos mean people may chose to ignore messages of 'don't touch' and risk their lives in pursuit of small financial gains.

Sources and further reading: UXO Lao website www.uxolao.org

the communist Khmer Rouge come to power in Cambodia for a disastrous period from 1975–1978, before invading the country and ruling it until 1989. Under the leadership of Pol Pot the Khmer Rouge sought to create a radical rural socialist utopia by emptying the cities of their populations, executing intellectuals, abolishing religion and family life and replacing them with agricultural communes. The experiment failed horribly, with starvation, disease and executions causing 1.2–2 million Cambodians, almost a third of the population, to lose their lives during Pol Pot's brief but devastating rule (see Sodhy 2004). Cambodia has since rejected socialist models and, as explained in Chapter 3, Vietnam and Laos have become more open to market-oriented approaches.

The ASEAN-5 countries that sided with the US experienced a much more peaceful period of political and economic development. In return for adopting the basic capitalist tenets of modernisation theory as their guiding development philosophy these states received considerable amounts of aid, FDI and political support from Western countries. The main requirement from the West was simply that these nations clamp down on communist movements, something that met the global geopolitical goals of the US but not something that necessarily accords with the participatory principles of good governance. The ASEAN-5 recognised their shared goals and concerns to form the Association of Southeast Asian Nations in 1967. ASEAN provided a forum where they could share their interests about internal and external communist influences; expectations of economic aid from the US, Britain and Japan; and hopes of expanding and revitalising regional cooperation post-*Konfrontasi* (Hagiwara 1973). While ASEAN initially formed in opposition to communism it has since grown to incorporate all Southeast Asian nations, except Timor-Leste which has observer status, and is now the most important political institution in Southeast Asia, contributing regional ways of pursuing political development.

In post-Cold War Southeast Asia international development institutions have become increasingly influential in domestic policy. The Asian economic crisis enhanced the power and influence of development organisations such as the IMF and the World Bank which insisted on economic and political reforms as a condition of lending. This included pressure to adopt political reforms based on neo-liberal concepts of 'good governance' that involve more transparent and accountable governance systems as they are considered more responsive to the needs of people, investors and markets. This has encouraged democratic governance systems as well as the decentralisation of administrative and political authority to encourage greater accountability. Grassroots NGOs, who also grew in power and popularity as a result of the economic crisis, also favour decentralised democratic systems of governance to empower ordinary people to participate within political processes. Democracy has since spread to Indonesia and decentralisation programmes are being implemented there as well as in Thailand and the Philippines. The newly independent states of Cambodia and Timor-Leste, both of whom were heavily reliant upon international development institutions when rebuilding, are also pursuing democracy and decentralisation. Similarly the communist states of Vietnam and Laos have been encouraged and are slowly exploring political decentralisation as part of their economic reforms, although open democracies are unlikely for some time yet. International development agencies have been less influential in places such as Myanmar, which has pursued an isolationist development pathway, or in Singapore, Malaysia and Brunei, all of which have been less reliant of foreign development institutions because of their strong domestic economies. In each of the countries authoritarian or semi-democratic political systems have been retained.

The influence of foreign institutions upon the political development of Southeast Asian states raises the question of sovereignty. Southeast Asian countries that are particularly reliant upon foreign institutions are sometimes referred to as aid-dependent nations as they have limited capacity to independently pursue their own forms of development. One way of conceptualising the dynamic between international institutions and state autonomy is to adopt the concept of graduated sovereignty. This can be used to refer to the different degrees of autonomy that the state has over guiding different sectors of society, as well as different spaces and communities within its territory. During IMF reforms in Indonesia, for example, the state may have had complete sovereignty over policing and justice systems but diminishing autonomy when it came to macro-economic decisions. In the same way when international NGOs work in Southeast Asia they may take on some of the functions of the state by providing medical services or contributing to education systems. The state then loses some control over these systems in particular spaces and cedes some authority to the international agencies working there. Hence sovereignty is not absolute throughout the country, instead it is graduated according to sector, place and international influence.

International NGOs are particularly influential in rural and remote communities in poorer Southeast Asian countries where they provide services in place of the state. In these cases NGOs commonly encourage grassroots community scale decision-making that employs participatory techniques and encourages all members of the community to decide on local development initiatives. These initiatives can vary from the provision of housing, water, sanitation, roads or equipment, to education, training courses, health programmes and human rights awareness campaigns. In most places NGOs are encouraged by states as they bring resources and skills that can assist in national development strategies. However, some states, such as the military government in Myanmar, feel threatened by their loss of sovereignty and monitor NGO motives and activities carefully. In places where the state is weak, such as Timor-Leste, NGOs can play a much more important role in community development than the government. While this can lead to unevenness between different communities it is also likely to encourage the development of more open governance systems as communities expect more say in planning decisions that affect their futures.

Internal influences upon political development

The political development of Southeast Asia reflects internal pressures as much as external ones. While states were trying to negotiate their role within the Cold War geopolitical order they were also trying to negotiate authority within their own territorial borders. One of the most common challenges derives from the multi-ethnic make up of post-colonial territories (see Table 4.1). Lande (1999) has classed ethnic minority conflict in Southeast Asian countries in four ways:

- Lowland versus highland conflicts.
- Archipelagic conflicts as diverse ethnic groups seek independent states.
- Conflicts derived from colonial borders that do not coincide with the cultural boundaries of ethnic groups.

Table 4.1 ***Ethnic diversity in ASEAN nations***

Country	*Main ethnic group*	*EM/IP population*	*Key EMs/IPs*
Brunei	Malays	21%	Chinese, indigenous groups
Cambodia	Khmers	6%	Cham, Chinese, Vietnamese
Indonesia	Javanese	50%	Sundarese, Madurese, Minang, Batak, Buginese, Balinese, Chinese, Banjarese, West Papuan, South Moluccan, Gayo, Alas, Minahasans, Dayaks
Laos	Lao	40%	Phuthai, Khamu, Hmong, Lue, Chinese, Vietnamese
Malaysia	Malays	42%	Chinese, Indian, Iban, Bidayuh, Melanau, Orang Asli
Myanmar	Bamars/Burmans	34–53% (estimate)	Karen, Shan, Chin, Mon, Arkanese, Kachin, Wa, Chinese, Tamils
Philippines	Filipinos	10.5%	Moro, Lumad, Igorot, Caraballo, Negrito, Mangyan, Palawan
Singapore	Chinese	21%	Malays, Indians
Thailand	Thai/Tai	2.3%	Malays, Karen, Hmong, Lahu
Vietnam	Kinh	14%	Chinese, Tay, Tai, Muong, Khmer, Hoa, Nung, Hmong, Mien, Gia Rai, Ede

Source: Clarke (2001)

Note: EMs/IPs refer to ethnic minorities/indigenous peoples.

- Conflicts between native inhabitants and more recent immigrant populations.

Lowland versus highland conflicts derive from pre-colonial times when highland people had little to do with lowlanders, developed autonomous identities and subsequently resented their post-colonial inclusion in states dominated by lowland ethnicities. Such conflicts are common across mainland Southeast Asia where minority groups such as the Hmong, Karen, Shan and Mon people feel oppressed by lowland governance systems and attempt to defend their homelands and identities. Archipelagic conflicts are similar in that minority groups seek autonomy from states whose centres of governance are located on different islands and dominated by people of different cultures and ethnicities. Three Indonesian hotspots for independence activity have been Aceh in northern Sumatra, Irian Jaya and East Timor while Filipino conflicts have centred upon the southern islands around

Mindanao. Conflicts based on borders occur where people of a particular ethnicity have been cut off from their homelands. In the southern provinces of Thailand, for example, there are 1.5 million inhabitants that have far more in common with the peoples of Peninsula Malaysia, in terms of language, culture and religion, than with the Buddhists of Thailand. In recent years violence has erupted here with over 2,000 deaths. Indigenous–immigrant conflict has usually been directed at economically successful Chinese immigrants, particularly in Malaysia and Indonesia, although other incidents of inter-ethnic violence have shaken parts of Indonesia (see Figure 4.3).

In addition to ethnic unrest internal challenges, particularly during the Cold War period, have been associated with political challenges emerging from rural communist movements. Many of these movements found support from China or the USSR and posed real threats to the authority of non-communist states. Communism appealed to rural Southeast Asians because of its core philosophy of distributing wealth more equally than what appeared to be occurring

Figure 4.3 Selected conflict points in contemporary Southeast Asia.

under capitalism. With rural spaces generally much poorer than urban spaces and more distant from the governing and policing apparatuses of the state, the conditions were positive for communist organising. The Pathet Lao, which secured power in Laos, has its roots in rural communist organising as did the Khmer Rouge in Cambodia. However communist movements were much more widespread with Thailand, Malaysia, the Philippines, Burma and Indonesia, all experiencing considerable challenges to their authority. Militant movements often resided in remote forested areas where they were hard to track down but able to seek rural support while launching attacks on government targets.

The response of Southeast Asian states to these challenges has shaped their political development. The responses employed can be classified as repressive, attractive or accommodating. Repressive approaches have epitomised the Myanmar government's responses to ethnically distinct highlander demands for autonomy as well as Indonesia's response to communist and ethnic insurgencies, particularly during Suharto's 31-year rule (1967–1998). In both countries the state has empowered the military to launch violent offensives against oppositional groups. Hundreds of thousands have fled Burma and sought refugee status in Thailand while many more have been killed inside the country. In Indonesia between 500,000 and 1.5 million communists are thought to have been killed when Suharto came to power and asserted his authority (Kingsbury 2005). Smaller scale repressive measures can be seen in Laos, Vietnam and Thailand where minority groups report persecution at the hands of the state. In Laos, for example, the Hmong have been treated badly for siding with the US during the 'secret war' and over 300,000 have either fled the country or been forced into re-education camps. These types of repressive measures have sought to force minority groups to accept the authority and power of the state.

The second response has been to try and minimise ethnic differences by attracting people to the benefits of a united nation-state. All Southeast Asian states have undertaken nation-building exercises in attempts to develop a popular and unified national identity. Vietnam and Laos, for example, have emphasised comradeship through communist ideologies and stressed the importance of loyalty to the Communist Party; Malaysia and Brunei have emphasised their Islamic religious identities and made Islam the state religion; the Philippines prides itself upon democracy and Catholicism; and Thailand promotes respect and reverence of the monarchy. Indonesia has launched a

The five principles of Indonesia's *Pancasila*:

1 Belief in one supreme God (monotheism)

2 Just and civilised humanity (humanitarianism)

3 The unity of Indonesia (nationalism)

4 Democracy through representative deliberations (consultative democracy)

5 Social justice for all Indonesians (social justice)

Figure 4.4 Principles of *Pancasila*.

unique nation-building strategy based on the five principles of *Pancasila* (see Figure 4.4). These principles have been taught at schools and emphasised at politically symbolic events to build the foundations of a united national identity within a diverse state. Other nation-building strategies have involved the expansion of state planning and services to more marginal areas to attract their support. Conscious efforts have been made to bring schools and health services to rural and remote spaces, as well as policing and surveillance mechanisms, as a means of promoting national ideologies while simultaneously diminishing rural discontent. The ultimate aim of such approaches is to create national identities that surpass ethnic and political differences.

The third approach, which would seem to have the most in common with the equitable development perspective given at the start of this chapter, is to accommodate for the concerns of minority groups, experiment with power-sharing arrangements and explore options regarding autonomy. While this has been a rare response in Southeast Asia it is gradually becoming more popular as democracy spreads through the region. In the Philippines, for example, the Moro people in the south of the archipelago have been granted a degree of self-governance through the establishment of the autonomous region of Muslim Mindanao. Similarly the province of Aceh in Indonesia has also been granted political autonomy over most areas of governance apart from foreign affairs, defence and fiscal policy. This is a substantial shift in state policies in both countries that were previously focused on repressive measures. These approaches provide hope that they may be extended to other parts of the region, particularly places like Indonesia's Irian Jaya which still suffers extreme forms of repression (see Box 4.2).

Box 4.2

West Papua: the struggle for independence

In 1945, at the close of the Second World War, the leader of the Indonesian Nationalist Party, Sukarno, unilaterally declared independence. The Dutch, intent on keeping their colonial empire intact, entered into a sporadic but bloody conflict with the emerging nationalist Indonesians but eventually ceded control, under much American and United Nations pressure, in 1949. The Indonesians claimed that all Dutch possessions should now become part of the new Indonesian Republic, including Dutch New Guinea, which had belatedly been acquired by the Netherlands in the 1890s. The Netherlands, however, believed Dutch New Guinea should be independent and invited the Indonesians, who refused, to take their claim to the International Court of Law.

In April 1961, a Papuan parliament overseen by the Dutch was elected to steer the colony to independence by 1970. The parliament erected a new flag, the Morning Star, on the 1 December 1961 and claimed West Papua as its new name (see Figure 4.1). This infuriated the Indonesians who immediately launched an unsuccessful airborne invasion. Pressure on Holland increased during the Cold War as the US and Australia both expressed support for Indonesia who they valued as an important anti-communist ally. To resolve tensions a vote was organised in which Papuans could decide on whether they wanted to integrate with Indonesia or form an independent West Papua. Amidst a climate of Indonesian intimidation and fear only 1,022 people voted and the pro-Indonesia result is considered to be unrepresentative of the wider population. West Papua was formally annexed by Indonesia in August 1969 and renamed Irian Jaya in 1973.

The incorporation of West Papua into Indonesia forced local people to cede development authority and power to a Java-based government and people they knew little about. The Indonesians have tried to develop a sense of national identity among Papuans by bringing in Bahasa Indonesia as the official language and encouraging the propagation of Pancasila principles. In addition approximately 750,000 migrants from densely populated areas of Java have moved to the province under the country's controversial transmigration programme (see Chapter 7). However resistance to Indonesian rule is still common with the huge Freeport gold and copper mine, which has generated huge profits for the American corporation as well as the Indonesian government, the centre of much frustration. Papuans, who have seen relatively little financial gain or economic development from the mine, have experienced widespread downstream environmental degradation (see Chapter 8).

West Papua still seeks the right to direct its own development but unlike Aceh, which has negotiated an autonomy package, and Timor-Leste, which is now fully independent, true autonomy for the province is far from guaranteed.

Instead there are a series of reports that claim serious human rights abuses, including executions, torture and rape, have been inflicted upon Papuan people. The decision by Australia to grant 43 West Papuans temporary refugee status in 2006 shows how seriously these claims are taken. On the 1 December each year West Papuans display the Morning Star flag in defiance of Indonesian authorities to symbolise their desire to pursue an independent future.

Sources and further reading: King (2004), Bertrand (2004)

The culture of political development

In this book political development has been presented as a process that leads to greater political freedoms among the populace and enhances opportunities for citizens to participate in governance. However such a definition is not universally agreed upon and it has received some concerted opposition from leaders from two of Southeast Asia's most economically developed countries. Prime Ministers Lee Kuan Yew of Singapore (1959–1990) and Dr Mahathir Mohammad of Malaysia (1981–2003) have been the two most prominent advocates of an alternative view, commonly known as the Asian values argument. Other leaders such as Suharto of Indonesia, the Sultan of Brunei and Burma's military leaders have also supported the Asian values perspective at various times. The Asian values argument suggests that there are particular Asian traits, ideas and values, often loosely linked to Confucian values inherited from China, that differ from the values prized by Western nations. Asian values are typically portrayed as favouring family, community and nation over the individual; the privileging of harmony and stability over individual rights; a recognition of the centrality of religion to life and happiness; an emphasis on savings, thrift and hard work; and respect for political leadership (see Milner 1999). Asians, it is argued, are different to Westerners, and therefore should pursue forms of political development that are better suited to Asian cultures and societies. A subset of this argument suggests that Westerners are overly preoccupied with civil and political rights, manifested in calls for more democratic systems, than social, cultural and economic rights, manifested in improved living standards and cultural integrity (see Table 4.2). Asian leaders argue that granting some civil and political rights, such as the right to vote or of free speech, can cause instability that detracts from the ability of the majority to realise more important social, cultural and economic rights (see Van Ness 1999). The most formal articulation of these concerns came in the 1993 Bangkok

Table 4.2 ***Categories of human rights***

Selected civil and political rights	*Selected social, cultural and economic rights*
Right to life	Right to work and earn wages that support a minimum standard of living
Rights barring slavery and torture	Right to equal pay for equal work
Right to equality before the law	Right to form trade unions
Right to privacy	Right to adequate standard of living (including food, clothing and housing)
Right to freedom of thought, conscience and religion	Right to highest standard of attainable health
Right to freedom of opinion and expression (media freedoms)	Right to education
Right to freedom of assembly and association	Right to take part in cultural life
Right to not be discriminated against	

Declaration on Human Rights in which Asian leaders affirmed the importance of social, cultural and economic rights, which they claimed were under-appreciated in most Western human rights debates. Under this argument Western calls for democracy are likened to a form of post-colonial imperialism unsuited to the Asian context, or what Aung-Thwin (2001: 494) has controversially called a democracy, '*jihad*, a holy war, backed by aggressive and confrontational rhetoric as well as sanctions or support'.

The Asian values argument had particular strength and conviction during the early 1990s when Asian economies were developing their economies rapidly and countries such as Singapore and Malaysia seemed to have identified a particular Asian way of successfully pursuing development. Since the economic crisis of 1997, however, Asian leaders have not talked about Asian values as frequently or with the same level of conviction. Not surprisingly the Asian values argument has met many critics both in the West and in Asia. A common criticism is that to talk of generic 'Asian values' is a misnomer as there is very little to link an Islamic Afghani at the western edge of Asia with a Buddhist Thai or Confucian Singaporean in Southeast Asia. Hence to talk of some form of collective Asian values is impossible given the vast geographic and cultural diversity across what is known as Asia (albeit a similar argument could possibly be mounted against universal human rights claims). A second criticism is that the Asian values argument has been developed by Asian leaders to legitimise their authoritarian style of governance. Civil libertarians in Southeast Asia suggest that such values are more a mythic invention of elites to maintain the status quo and their grip on power than a true reflection of the values of the masses. Indeed Hill's (2000) research suggests that Asian values are taught through government education systems rather than necessarily reflecting true indigenous household values. While there is probably truth in both of these critiques there can be no doubt that the Asian values argument has proved an interesting counterpoint to the liberal democratic obsession of the West and raises questions about core development assumptions. Could the enormous social and economic gains experienced in Singapore and Malaysia have been achieved in a less stable but more democratic political environment? And, what is more important, the improved health and education systems now available in these countries, or the right of individuals to vote? It is an interesting argument that has been used to legitimise some of the more peculiar political systems currently present in Southeast Asia and has provided an impediment to the evolution of more inclusive forms of governance. What it also emphasises is the need to consider the multiple dimensions of development, that economic development does not necessarily imply political development, nor vice versa, instead different sectors of societies develop at different rates according to different socio-cultural, political and economic contexts.

Asian values have become associated with the ASEAN political structure. Since its early Cold War inception as an institution opposed to communism it has grown from its original members of Thailand, Malaysia, Indonesia, Singapore and the Philippines to incorporate all Southeast Asian countries apart from Timor-Leste, which has observer status and hopes to be admitted sometime in the future. Brunei became the sixth member of ASEAN upon its independence in 1984, while the collapse of the Soviet Union and end of the Cold War paved the way for Vietnam to join in 1995, Laos and Myanmar in 1997 and finally Cambodia in 1999. ASEAN has contributed to a particular political culture known informally as the ASEAN way. In contrast to the Asian values argument that grants political autonomy to Asian leaders to pursue whatever they like, the ASEAN way promotes consultation and consensus among member states. Most negotiations take place in private and if consensus cannot be reached ASEAN nations agree to disagree and pursue other less contentious issues. While admirable, Narine (1999: 360) has noted: 'ASEAN has promoted the art of conflict avoidance, but not conflict-resolution'. The ASEAN way is epitomised in ASEAN policies of non-interference, in which member states agree not to interfere with the domestic policies of other member states, thereby enhancing the prospects of regional harmony. This approach has contributed to lasting peace between ASEAN nations but has attracted criticism from international partners, many of whom believe ASEAN should be doing much more to bring about change in its more politically repressive states. International pressure has led to some cracks in ASEAN solidarity, particularly when ASEAN leaders publicly criticised Myanmar's decision to imprison pro-democracy leaders, and privately pressured the regime to decline the ASEAN chairmanship role it was scheduled to take up in 2006.

A further change to the culture of political development has occurred in the aftermath of the 11 September 2001 terrorist attacks on the US. The subsequent US-led 'war on terror' has sent shockwaves of the initial assault around the world. In Southeast Asia the war on terror has seen strengthened partnerships between ruling states and US-intelligence agencies, strengthening state capacity to monitor dissenting ethnic or political groups, many of whom have been discursively reclassified as 'terrorists'. The most prominent terrorist organisation in Southeast Asia is the Jemaah Islamiah (JI) network, which has been blamed for the bombing of churches and two Bali nightclubs in 2000, the bombing of the Marriot Hotel in 2003, and contributing to conflicts in Ambon, Sulawesi, the Philippines and elsewhere (Kingsbury 2005). JI members have subsequently been arrested in Malaysia, Singapore, the Philippines and Indonesia (Abuza 2002; Rahim 2003). On the pretence of uncovering this network Glassman (2005) has argued that state power has been reified in Southeast Asia slowing moves towards more open democratic societies. This can be seen in the repressive internal security laws that allow states to arrest and hold terrorist suspects without trial for extended periods of time. It can also be seen in the closer ties between the US and the Indonesian military, which legitimises this institution despite its questionable human rights record in Irian Jaya, Aceh and Timor-Leste. The new global geopolitics oriented around an elusive anti-terror concept may or may not reduce acts of violence within the region but appears likely to stifle progress towards more open forms of political development.

Political systems and political development

External influences, internal influences and the culture of political development in Southeast Asia have resulted in a diversity of political systems. Each can be assessed according to the principles of good governance and the participatory opportunities they provide for their citizens.

Decentralising democracies: Philippines, Thailand and Indonesia

The three states that are pursuing decentralised democracies are the Philippines, Thailand and post-Suharto Indonesia. While democracies pre-exist the Asian economic crisis in two of these countries it is interesting to note that each borrowed money from the IMF in the wake of the crisis and have since been encouraged to undergo substantial economic and political reforms. The longest standing democracy is in the Philippines which, apart from a 14-year period from 1972–1986 when the democratically elected President Ferdinand Marcos declared a state of emergency, has been ruled that way since independence. Marcos was eventually forced from office after a popular uprising, known as the People Power Revolution, after years of corruption, intimidation and mismanagement. Democratic voting systems, media freedoms (see Table 4.3) and the separation of courts and state have since been re-established, leading to a stable democratic political system. The main criticisms of the Filipino democracy are that it is dominated by entrenched landowning elites and that informal patron–client relationships survive beneath the democratic façade (see Putzel 1999). Patron–client relationships, which are common throughout Southeast Asia, are informal political relationships that

Table 4.3 ***Media freedoms in Southeast Asian countries***

Rank 2007 (1–195)	*Country*	*Rating (1–100)*	*Status*
90	Timor-Leste	42	Partly free
100	Philippines	46	Partly free
114	Indonesia	54	Partly free
122	Cambodia	58	Partly free
126	Thailand	59	Partly free
150	Malaysia	68	Not free
154	Singapore	69	Not free
166	Brunei Darussalam	76	Not free
170	Vietnam	77	Not free
176	Laos	81	Not free
191	Myanmar	96	Not free

Source: Freedom House (2007)

Note: Freedom House produces an annual index of media freedoms. A score of 1 equates to the highest level of media freedom, and 100 indicates the lowest level of media freedom.

have been inherited from pre-colonial times. A politician or powerful member of society acts as a patron to other members of society or to businesses and industries by representing their interests in elite decision-making circles. The patron usually negotiates better economic conditions or social services for their clients in return for their political support and a proportion of any profits generated (Lim and Stern 2002). Patron–client relationships enhance the power of elites and sustain political inequalities, partly explaining why the democracy of the Philippines has not led to decrease in inequality and landownership remains extremely uneven. These types of relationships have been criticised by a range of civil society institutions in the Philippines and it is only recently that left-leaning parties have even begun contesting democratic elections given what they perceive as unequal playing fields (see Quimpo 2005). Nevertheless the entrenched democracy of the Philippines provides regular opportunities for people to participate in governance.

Thailand is a constitutional monarchy where the king is officially the head of state but the day-to-day running is left to a democratically elected prime minister. During moments of crisis however the king can get involved and current King Bhumibol, who received the inaugural UNDP Human Development Lifetime Achievement Award in 2006, has intervened at critical moments in socially progressive ways. Thailand has democratically elected its prime minister since 1992, however a military coup in September 2006 banished billionaire Prime Minister Thaksin Shinawatra from the country. The military has claimed it was acting in the people's interests after widespread claims of high level corruption. They have promised a return to democracy in the near future but in the meantime some national scale civil and political rights are being suspended, leading to criticisms from local pro-democracy groups. Like the Philippines there is concern about patronage politics and that 'big person-little person' roles still dominate political life and power relations more generally (see Kingsbury 2001: 161). In such situations less powerful members of society are likely to acquiesce to current inequalities for fear of reprisals and a lack of patronage.

Indonesia's flirtation with open democracies has been much shorter. After 33 years of President Suharto's semi-democratic rule, the Asian economic crisis led to widespread protests and his eventual resignation in 1998. Many media freedoms were restored, new political parties have formed and the first truly free Indonesian elections were held in 1999 and again in 2004. After such a long period of intimidation the elections are generally considered to be remarkable successes, notable for their openness and lack of violence. However, substantial political challenges remain for the world's third largest democracy. Transparency International recently ranked Indonesia 130 of 158 countries in its annual corruption perceptions index in 2006 (Liddle and Mujani 2006; see Table 4.4). There is also concern about patronage politics and how much freedom the military, which was so powerful under General Suharto, will allow civilian governments.

In each of these countries, with the temporary exception of Thailand, citizens are experiencing greater political freedoms now than in the past, suggesting positive political development. Most people possess the right to vote, to form political parties, to contest elections, to criticise government and now have access to a more open media. Of particular significance has been the trend towards political decentralisation in each of these countries. The central state has devolved considerable financial resources, decision-making power and regulatory responsibilities to local authorities. Local governments are now responsible for issues relating to natural resource management, public works, public services (such as health and education) and social welfare, while the central state retains control of issues like defence, foreign affairs and justice. In the Philippines and Indonesia, and some parts of Thailand, authorities are elected by the local constituency and are seen as more accountable and responsible to local concerns. From the perspective of grassroots or post-development theory this is a positive step forward as localities now have more ability to pursue their

Table 4.4 ***Perceptions of corruption in Southeast Asia***

Rank 2006 (1–163)	*Country*	*CPI*
5	Singapore	9.4
44	Malaysia	5.0
63	Thailand	3.6
111	Laos	2.6
111	Timor-Leste	2.6
111	Vietnam	2.6
121	Philippines	2.5
130	Indonesia	2.4
151	Cambodia	2.1
160	Myanmar	1.9
n/a	Brunei Darussalam	n/a

Source: Transparency International (2006)

Note: The CPI (Corruption Perception Index) ranks countries by their perceived levels of corruption, as determined by expert assessments and opinion surveys. The lower the CPI the greater the perception of corruption.

own visions of development, rather than succumb to state-favoured models such as modernisation. However there are many challenges associated with decentralisation. Beard (2005), for example, has written on the difficulties of getting marginalised people and women to participate in decentralised government structures; Hadiz (2004) has questioned whether decentralisation shifts state corruption and uneven power relationships to the local scale; and Bebbington *et al.* (2006) have expressed concern over the limited capacity and skills of local scale villagers to resolve issues and pursue effective forms of development. Nevertheless in shrinking the gulf between citizens and what are now local decision-makers these three countries are progressing towards participatory forms of governance. Community concerns are more likely to be reflected in local governance structures than national ones and opportunities for political participation and empowerment are much higher in decentralised states than centralised ones. There are also greater opportunities for local communities to pursue their own forms of development.

Post-conflict democracies: Cambodia and Timor-Leste

Cambodia and Timor-Leste have both recently undergone the transition to democracy after years of internal strife and foreign occupation. Cambodia has the longer democratic history having its first United Nations-supervised elections in 1993, which resulted in an unlikely coalition government between Prince Norodom Ranariddh's FUNCINPEC party, who became Prime Minister One, and Hun Sen's Cambodian People's Party (CPP), who became Prime Minister Two. This unstable coalition broke down in 1997 when Hun Sen violently took power and has ruled the state ever since. International pressure led to elections in 1998 and 2003 but doubts have been raised about the validity of the electoral process with claims of voter intimidation by CPP-dominated police, arbitrary detention and killing of non-CPP activists and candidates, limited media access for non-CPP parties and the instigation of CPP-dominated judiciary. Within this unhealthy political climate in Cambodia international pressure has seen a decentralisation programme introduced and to that end commune elections took place in 2002. These local scale government bodies have limited resources but do provide limited opportunities for community empowerment. Current research, however, suggests that the country's history of violence, lack of educational infrastructure and entrenched patron–client relationships are preventing people from actively seeking out and engaging in these opportunities (Un 2006).

This may change in the future but until that time Cambodia is clearly lagging behind the more established democracies in terms of its political development.

Timor-Leste received a huge amount of aid when it secured independence from Indonesia and has subsequently pursued, with United Nations assistance, a democratic governance system. The first national scale elections were held in 2001 and saw the FRETLIN party, which was the core organisation resisting Indonesian occupation, winning 57 per cent of the vote. Local scale elections have also taken place and considerable responsibilities are devolved to this tier of government. International development agencies have been building the capacity of local governance structures by giving grants directly to village committees rather than central government and developing ongoing partnerships with local community counterparts (see Box 4.3). Like Cambodia however, Timor-Leste's history is clouded

Box 4.3

Development, dependency and community participation in Timor-Leste

Development in Timor-Leste has been strongly shaped by international development agencies since the ballot for independence in 1999. When the results of the vote were made public the country was engulfed in violence as pro-Indonesia militia went on a murderous rampage that destroyed lives, property, industries and agriculture. When the Indonesian administration withdrew the challenges for development were enormous with a new country having to be built from the rubble of the past. With few Timorese learning governance skills during Indonesian rule the United Nations stepped in and setup the United Nations Transitional Administration in East Timor (UNTAET) to police, oversee and design the development of the new state. During this period Timor-Leste was effectively run by the UNTAET, in consultation with local advisory groups, before officially declaring independence in 2002.

Since then a lack of domestic capacity has forced Timor-Leste to continue its heavy reliance on international institutions. Development organisations provide staff who advise or take up important positions within government and fund a wide range of projects ranging from road building and water supply to psycho-social counselling and women's refuges. In Dili a local economy has formed to support this industry where exclusive bars, cafés and internet sites cater for the needs of foreigners. Amidst this heavy international influx are concerns about the dependency of the Timorese state on foreign aid and the effects this may have on national development.

In line with more recent development theories many development organisations are experimenting with community-based development. These programmes include ongoing partnerships between local communities and international NGOs; council to council relationships through Australian Friends of Timor-Leste scheme; and small grants programmes where community-based organisations can approach international donors, rather than the state, for funding. It is hoped such initiatives will lead to independent grassroots development in which local communities are empowered to pursue their own development directions. The sensitivities of this approach are evident in the following reflections of an NGO worker working in a community partnership:

> There's one community that's in Oecussi and they have a lot of food taboos and one of them is no fish or seafood products because that's their ancestry. We just felt that was like completely taboo and was untouchable really, so we tried to work with them on other protein sources . . .
>
> (McGregor 2007: 165)

Things do not always run so smoothly however and uneven power relationships can lead to conflict:

> In Viqueque they posted 2 international staff for 6 months or 1 year . . . and they say 'you have to do it like this! In my country we do it like this! You have to do it like this!' Then one day they come to these internationals with machetes and they have to run away from this village.
>
> (McGregor 2007: 165)

Power imbalances can also affect small grants programmes where donors have the ultimate say over which proposals will be funded. In a disturbing misuse of power there is evidence that small grant funding that had been allocated by the Australian government to a local NGO was withdrawn when the NGO signed onto a petition opposing Australian policies regarding oil in the Timor Sea. These types of occurrences raise concerns that not only states, but communities, can also become aid-dependent.

Sources and further reading: McGregor (2007), La'o Hamutuk (2006)

in violence and its high levels of poverty detract from people's ability to engage in political processes. In 2006 community frustrations erupted in riots which eventually contributing to the resignation of Prime Minister Mari Alkitiri amidst claims of corruption and economic mismanagement. In this sometimes tense political climate, and in a country struggling with high levels of poverty, the long-term political development of Timor-Leste remains unclear.

Semi-democracies: Malaysia and Singapore

Case (2002) identifies Malaysia and Singapore as semi-democracies. This means that there are regular elections contested by a range of parties who are permitted to organise, operate headquarters, solicit contributions and select candidates and leaders. However, oppositional parties have difficulty accessing the broader public because the government controls most mainstream media outlets and they are restricted from circulating their own party publications or organising mass rallies – even during campaigning periods. Oppositional leaders who break these agreements can be imprisoned or sued. In controlling public information semi-democratic governments can promote the veneer of democratic process comfortable in the knowledge that their power is never likely to be seriously threatened. This gives them an aura of legitimacy both domestically and among international audiences, further heightened by their staunch defence of Asian values. International calls for greater democracy in Singapore and Malaysia, for example, are barely whispers when compared to the chorus of voices attacking Myanmar's overtly non-democratic regime.

Singapore has been ruled by the People's Action Party (PAP) since independence. Prime Minister Lee Kuan Yew ruled from 1959–1990, Gok Chok Tong followed until 2004 when has since been replaced by Yew's son, Lee Hsien Loong. The longevity and success of PAP was, perhaps, foreseen by Yew as early as 1960 when he was quoted in a radio interview as saying: 'If I were in authority in Singapore indefinitely without having to ask those who are governed whether they like what is being done, then I have not the slightest doubt that I could govern much more effectively in their interests' (Kingsbury 2001: 337). Some of the political tactics of PAP have included: gerrymandering of electoral boundaries so that areas voting for opposition parties are incorporated into zones that are overwhelmingly pro-government to ensure PAP success; the formation of pro-PAP Residents Committees that receive preferential government services; controlling media content and restricting oppositional advertising and promotion; and launching high profile defamation cases that reflect close relationships between judicial and political systems (see Kingsbury 2001). More recently there has been censorship of online blogs and films, including an Amnesty International speech against the death penalty (see Rodan 2006). Despite this level of control the population is not overly convinced of PAP credentials – in the 2006 election PAP won 82 of 84 seats but only 66 per cent of the overall vote. Calls for political reform remain muted, however, with PAP's governance overseeing a period of unparalleled economic success.

Malaysia has also been ruled by a single party, the United Malays National Organisation (UNMO), since independence. UNMO's support has been falling in recent years, only receiving 35.9 per cent of the vote in the 2004 elections, but it has been able to maintain its majority through coalitions with the Malaysian Chinese Association (15.5 per cent) and a variety of minor parties mostly based in Borneo. The coalition is known as *Barisan Nasional* (National Front) and was led most famously by Prime Minister Mahathir Mohammed from 1981–2003. UNMO has maintained its grip on power through successful economic development strategies, clever coalitions with minor parties and through curtailing particular civil and political rights. UNMO, like PAP, is accused of gerrymandering, stacking the courts with sympathetic judges, controlling the media and suing those who post critical views (see Kingsbury 2001). However the recent electoral results suggest that oppositional parties are managing to boost their presence and a more open political system may be evolving. Many hope the ascension of the more liberal Abdullah Ahmed Badawi to the position of prime minister in place of Mahathir will initiate a new phase of political development.

In Singapore, which is a small city-state, and Malaysia there have been few moves towards decentralizing power. Instead leaders of both countries have championed the Asian values argument and have backed up their rhetoric by overseeing periods of impressive economic growth. While both countries were hit by the Asian economic crisis it did not lead to the type of political crisis that engulfed Indonesia and saw the resignation of Suharto. Instead Mahathir drew on nationalist rhetoric to blame international currency speculators for Asian troubles, particularly targeting Hungarian-born billionaire George Soros who he blamed for initiating the rush to withdraw FDI. As a consequence there are few opportunities for people to involve themselves in governance or pursue non-state approved development pathways. Instead the state retains control of development directions and allows very little dissent from official policies. While most people tacitly agree to this, as a compromise for the economic wealth this political model has delivered, it is a source of frustration for poor, marginal and disempowered groups who may struggle, or risk imprisonment, to have their voices heard. This applies to factory workers who suffer from a lack of effective union representation as well as minorities who must succumb to national norms rather than their own cultural preferences (see Chapter 5).

Socialist states

The economic reforms taking place in Vietnam and Laos have not been matched by the same rate of political reform. Changes are occurring but the basic communist political structures remain intact. There are no openings for alternative political parties, the media is still heavily controlled by the state and positions of authority are filled by Party members. While both countries are still highly centralised in terms of power and decision-making there have been moves towards decentralisation, particularly in Vietnam, where local authorities now have more say in the management of economic and social infrastructure, yet there is a long way to go. In theory the Communist Party is the people's party and participatory decision-making processes at the local scale negate the need for formal political democracies or oppositional political parties. However historically suspicious relationships between citizens and the state have seen people persecuted for airing alternative ideologies and limited the participatory potential of these governance systems. This is particularly problematic for minorities who may feel their views are not effectively catered for by central authorities. Nevertheless there is greater openness about the Vietnamese government in recent years, evident in the flow of migrants back to the country and in its recent white paper on human rights that guarantees, among other things, greater freedom of the press (Luong 2006). Laos lags behind, however, with civil society continuing to be repressed through bans on non-state NGOs (only the bizarrely named GONGOs – government organised non-governmental organisations are permitted to exist) and strict limits placed on media expression (Forbes and Cutler 2006). Political development in both countries appears to be lagging behind their recent economic reforms.

Authoritarian states

The two remaining states of Southeast Asia are variants of authoritarian governance systems. The tiny country of Brunei Darussalam has been ruled by Sultan Hassanal Bolkiah, who shares the titles of prime minister, minister of defence and minister of finance, since 1967, although it has only been fully independent of Britain since 1984. The sultan's power has been secured since the declaration of emergency powers in 1962 which have never been revoked. Despite the authoritarian nature of his rule the Sultan is considered by many to be a benign ruler who has provided free education, health care, subsidised food, fuel and housing, and low interest loans for government employees, all financed through the booming oil industry. Nevertheless the media has little freedom of speech, many outlets being owned by the sultan's family, and there is little political space for oppositional parties to organise and contest the sultan's rule. In many ways the successful economic development of Brunei has retarded political development in a manner that mirrors Malaysia and Singapore; people have traded off the right to vote and participate in governance in return for economic progress.

Myanmar, in contrast, is roundly regarded as one of the most abusive authoritarian states in the world. The current military government came to power through a coup in 1962 when it replaced a democratically elected government that had ruled since independence. The State Law and Order Restoration Committee (SLORC) has fought long violent wars with ethnic minority groups, recruited an extensive informant network to monitor people's behaviour, engaged in forced labour, restricted media freedoms, and imprisoned thousands of political protestors. The most famous political prisoner is Aung San Suu Kyi, daughter of assassinated independence leader General Aung San, and head of the main opposition party, the National League for Democracy (NLD). In 1988 widespread protests threatened the authority of SLORC who responded with a violent military crackdown, a name change to the State Peace and Development Council (after consultations with an American PR firm) and the announcement of national elections in 1990. The NLD won by a landslide however the results have never been officially recognised by SLORC and Suu Kyi has been under house arrest, more or less, ever since. Despite immense international pressure from United Nations organisations, aid agencies, the US government and international activist organisations such as Amnesty International there have been few indications of political reform within the country (McGregor 2005). There are no opportunities for people to influence the military governance apparatus except through warfare, as many ethnic minority groups have experienced for decades. Unlike the semi-democracies of Singapore and Malaysia or authoritarian Brunei, Myanmar's military has not been able to bring economic prosperity to the nation and domestic frustration with the regime is high. More open and participatory governance structures are a long way off in a country that is still ruled through intimidation and fear.

Conclusions

There has been a general trend towards more inclusive and participatory systems of governance within the region. This is largely a reflection of Indonesia's transition to democracy as well as flirtations with democracies in the smaller countries of Timor-Leste and Cambodia. These countries, as well as Thailand and the Philippines, are also beginning to explore decentralised decision-making structures, providing greater autonomy to local authorities. The evolution of semi-autonomous areas, as well as the succession of Timor-Leste from Indonesia, suggests that central governments are becoming more open to power sharing and accommodating minority interests. In this sense political development has been positive as there are more freedoms and opportunities for local people to engage in political processes. In contrast, however, leaders of wealthier economies have showed little interest in decentralisation or democracy. Instead they have used their credentials as good economic managers and the Asian values argument to legitimise their repression of more open forms of political development. This appears to have appeased the majority of their populations and silenced pro-democracy movements who may fear persecution for speaking out. Either way the economic success of these nations challenges the common assumptions that free and open political democracies are important elements of economic growth. Elsewhere political development has been repressed by governments who believe the inclusiveness of socialist systems require little reform, or, as in the case of Myanmar, have conducted military and police actions against their own populations to prevent reforms. This enormous array of approaches to political development shows no sign of narrowing, indeed the non-interference principles of ASEAN ensure the political development pathways of the region will retain their diverse and idiosyncratic natures for years to come.

Summary

- **The principles guiding political development are based on the concept of good governance; however this concept is contested and interpreted differently by different institutions.**
- **There is a wide range of political systems in Southeast Asia, which reflects both internal and external pressures on development.**

- **States have experimented with a range of nation-building exercises, which include repressive, attractive and accommodating stances, to retain control of their territory among fragmentary pressures.**
- **There is a gradual shift towards democracy and decentralisation in some countries that is opening up new opportunities for political participation and engagement with development processes.**
- **In other countries socialist ideologies and Asian values arguments, which champion social, cultural and economic rights over civil and political rights, are being used to maintain the status quo.**

Discussion questions

1 Discuss the different impacts external and internal forces have had upon the political development of Southeast Asian states.
2 Consider whether the relative economic successes of Singapore and Malaysia validate the Asian values argument.
3 Discuss the pros and cons of political decentralisation and how this fits the principles of good governance.

Further reading

For contemporary overviews of politics in Southeast Asia see:

Kingsbury, D. (2005) *South-East Asia: A Political Profile (2nd Edition)*. Melbourne: Oxford University Press.

Dosch, J. (2007) *The Changing Dynamics of Southeast Asian Politics*. Boulder, CO: Lynne Reinner.

For colonial and post-colonial conflicts see:

Christie, C. (2001) *Ideology and Revolution in Southeast Asia 1900–1980*. Richmond: Curzon.

Snitwongse, K, Thompson, W. (2005) *Ethnic Conflicts in Southeast Asia*. Singapore: ISEAS Publications.

For a detailed introduction to the different political parties and processes in the region see:

Leifer, M. (2001) *Dictionary of the Modern Politics of South-East Asia (3rd Edition)*. London: Routledge.

For an interesting take on the Asian values argument see:

Van Ness, P. (ed.) *Debating Human Rights: Critical Essays from the United States and Asia*. London: Routledge

Useful websites

Useful websites of NGOs that provide updates on human rights, political freedoms and media freedoms include:
Freedom House www.freedomhouse.org
Amnesty International www.amnesty.org
Human Rights Watch www.hrw.org
Also political developments can be tracked at the ASEAN site: www.aseansec.org

5 Social development

Introduction

Development is often conceptualised in economic or political terms but it also shapes and is shaped by social structures, beliefs and systems. Indeed it is often concerns about negative social change that is at the core of equitable development critiques. For the purposes of this book social spaces will be conceptualised as those spaces that exist between the state, the market and the household in which people of similar identities, positionalities or interests organise to resist or influence development processes. Hence the focus is primarily upon civil society, the collection of social movements and NGOs that represent social concerns and lobby for changes to development processes. From the perspective of equitable development civil society plays a vital role. It provides an alternative means through which people can be empowered to participate in development; it facilitates social networks in which people can imagine alternative futures; it strengthens people's capacity to contest the distribution of wealth and resources; and it allows people to lobby for more sustainable development approaches. Within alternative and post-development theories the activities that take place in social spaces are seen as the most transparent and organic representations of the people's needs, desires and concerns. This chapter will explore social spaces in Southeast Asia by introducing key elements of civil society before profiling social action relating class, religion and gender.

Civil society, social movements and development

Some of the most influential forces within civil society are social movements. Social movements are broad collections of people who mobilise around particular identities, such as working-class or farming identities, or particular goals, such as improving access to health care or overcoming discrimination. They seek positive change within society by targeting state, market or broader developmental processes, and lobbying for improvements to the status quo. Whatever the differences between these movements and the contexts from which they emerge, they are intimately concerned with development, addressing struggles over the allocation of resources, self-determination and rights of economic and cultural survival (Routledge 1999). There is a range of both formal and informal tactics that social movements adopt to achieve their ends. Formal tactics include strikes, petitions, letter-writing campaigns and lobbying politicians, while informal or unconventional tactics include protest marches, song and dance, blockades or 'everyday resistance' where workers may slow down their production, take sick days or engage in graffiti or vandalism. Independence movements were one of the earliest social movements to form in Southeast Asia but these have since been replaced by movements focusing on a wide variety of issues including ethnic, religious and gender discrimination, environmental degradation, governance, human rights and justice (Guan 2005).

An important component of social movements, but also entities that exist independent of social movements, are civil society organisations, the most common form within development being NGOs. NGOs are issue-driven not-for-profit institutions, distinguishing them from private or commercial entities, and, as the name suggests, are separate to, although not necessarily independent of, the state. There is now a broad array of NGOs in Southeast Asia mobilising around issues such as human rights, gender, labour, environmental degradation, HIV/AIDS and poverty, as well as actively providing important services such as shelter for the homeless or food for the poor. Such institutions vary in size from a couple of part-time volunteers concentrating on a particular local issue to large professional international development institutions such as Oxfam or World Vision that engage in a wide variety of programmes. NGOs are becoming increasingly important actors in Southeast Asian development for the following reasons:

- the growing democratisation of the region that has opened up new spaces for NGOs to organise and contest or engage with development processes;
- the adoption of neo-liberal economic theories, particularly in the wake of the Asian economic crisis, which has shrunk government services and necessitated the rise of new organisations to fill the gaps;
- concerns about good governance and state corruption that has seen some international donors prefer to fund local NGOs over government counterparts;
- the growing popularity of grassroots and post-development theories has seen international development institutions, particularly international NGOs, support domestic civil society organisations rather than the state.

While the strength of civil society differs from place to place NGOs are becoming vocal and important actors shaping and contesting Southeast Asian development.

The countries with the most active and diverse civil societies are generally those that have the longest history of democracy. Hence the Philippines, Thailand and, more recently, Indonesia, have well-established civil society movements that use democratic spaces to organise, participate within and contest development processes. Cambodia and Timor-Leste have much more youthful movements while the semi-democracies of Malaysia and Singapore encourage particular types of social activism, such as those focused on welfare and philanthropy, but restrict activities that may threaten state harmony or authority (Koh and Ling 2004). In Vietnam *doi moi* policies have led to a contraction of state services and a rapid rise of independent voluntary associations and international NGOs to fill the gaps (Luong 2005). Working in parallel with these new organisations are older GONGOs such as the Vietnam Women's Union and the Vietnam Youth Union. Such organisations are differently positioned to Western civil society organisations as they are structured to operate within the state, rather than in opposition to the state. Hence the Vietnam Youth Union is a state institution but at the same time it attempts to effectively represent young people's issues in state planning and decision-making. While some would argue that such an institution does not fit the definition of civil society as it is part of the state, very few NGOs exist without some sort of state support. This is particularly the case in developing countries where estimates suggest 80–95 per cent of NGO funding is derived from (mainly foreign) government sources (Weller 2005). It also reflects Asian traditions such as Confucianism where societies and states have traditionally worked cooperatively rather than independently or in opposition to one another. Civil society organisations are most constrained in Laos and Myanmar where GONGOs are the only officially recognised forms of domestic civil society permitted by the state.

The relationship between civil society and equitable development is a complex one. Civil society is necessary for ordinary people to become involved in development by empowering them to contest and engage within development processes. However there is little evidence in Southeast Asia that active civil societies lead to more equitable distributions of wealth and resources. In the Philippines, for example, a well-established civil society has not prevented the country become one of the most economically unequal within the region. Instead the wealthy but only semi-democratic state of Singapore has been much more effective in distributing resources throughout its economy, although some still miss out, than countries where civil society is freer to critique and contest government policies and processes. Communist countries also contest the relevance of civil society given their inclusive political ideals that theoretically make civil society obsolete. In other words free and active civil societies accord with the themes of empowerment and participation but do not guarantee more even forms of development. The influence of civil society upon development will now be examined in issues relating to class, religion and gender.

Class and development

Economic development in Southeast Asia has benefited some groups in society more than others. As discussed in Chapter 3 the overall improvement in wages and wealth has been accompanied by growing rates of inequality, which, in turn, is leading to an increasing stratification of Southeast Asian societies along class divides. While inequality has always been present within the region, historically a small group of powerful elites ruled over large numbers of agricultural producers, the pattern of inequality is changing. The most significant transformation has been the growth of large urban working and middle classes, particularly in the wealthier original ASEAN-5 economies that have benefited the most from rapid economic development (Hersh 1997). Class stratification emerged as secondary and tertiary industries formed that opened up a diversity of new income earning opportunities. The working class is comprised of the legions of low income factory workers that have underpinned urban economic development. The middle class is comprised of workers in higher paying occupations, such as managers, professionals and technical experts who supervise, advise and improve industrial practices. These more specialised positions have attracted higher wages and facilitated the rise of associated service industries in fashion, tourism and hospitality to cater for middle-class interests. These new classes have transformed the structure of traditional Southeast Asian societies with emerging class interests shaping civil society actions.

It would normally be expected that the emergence of distinct social classes would be associated with the rise of social movements, NGOs and labour unions to represent their interests in broader society. In Southeast Asia, however, this is not always the case. The most important institution for working-class interests are labour unions and labour NGOs that act to collectively represent worker interests in negotiations with employers and the state. Unions generally try to improve the wages, working conditions and safety standards of their workers through negotiations and actions such as strikes, 'work-to-rule' tactics and pickets. From an equitable development perspective labour unions are important institutions for mobilising workers and facilitating more even distributions of wealth and resources. For a variety of reasons labour unions have struggled to form and effectively represent worker interests throughout the region. At the national scale the ASEAN-5 states have traditionally been suspicious of labour unions, partly due to concerns about links with communist parties, but mainly because of the negative ramifications effective unionisation may have for their EOI/FDI economic development strategies (see Chapter 3). If unions secure higher wages and better conditions for their workers this will inevitably increase the costs of producing goods and make the country less desirable for foreign investment. In other words a cheap, un-unionised workforce has been an important driver of Southeast Asia's FDI-led economic development. As a consequence the ASEAN-5 states have made unionisation difficult and there are few truly independent or radical trade unions within the region.

Unions have been most free to organise in the larger democracies of the Philippines, Thailand and post-Suharto Indonesia. However they have been tightly controlled in places such as Singapore and Malaysia where most independent unions are either banned, as is the case for Malaysia's important electronics sector, or have strong affiliations with government. Those that fall out of line risk intimidation by the state, as has occurred in Malaysia where union leaders have been threatened under the anti-terrorist Internal Security Act as national security 'threats' (Kelly 2002). Labour unions in the communist countries of Southeast Asia are also strongly affiliated with the state. While this may have sufficed in the past due to socialist approval of workers rights, the move towards private FDI-led development risks a divergence of interests between the state, which is seeking to attract foreign capital by providing a cheap workforce, and the workers who want an equitable return for their labour. In this situation state-organised unions may not by effective in representing worker interests (Noerlund 1997). Independent union activities are even more limited within authoritarian states.

At a more local scale union activities have been slowed by industrial tactics adopted by employers. Some tactics include the set up of close links between managers and employees that encourage disputes to be resolved on an individual basis rather than collectively; the restriction of labour union representation to individual firms that allows internal resolution rather than industry-wide representation; preventing non-employees, such as trade union leaders, from accessing industrial estates; directing informal threats, pressure and even acts of violence towards union activists; and selectively employing people who are less likely to contest work conditions, such as migrants and women (Kelly 2002). Migrants, who may be domestic or foreign, are sought after as they are often contracted to the workplace and therefore unable to shift jobs irrespective of the conditions, and are usually hard workers as they have made considerable personal sacrifices to take up the job and have often left families at home who are reliant upon their incomes. Women are also preferred by employers for many of the most monotonous and lowest paid jobs in textile, footwear and electronics industries. This is due, in part, to the gendered assumption that they are more likely to passively accept low wages and poor working conditions than their male counterparts. These approaches by states and employers have suppressed the evolution of more effective labour unions to represent worker interests and mobilise for more equitable development patterns.

The emergence of a large middle class, which is particularly pronounced in the wealthier nations, also has ramifications for civil society. The middle class is generally wealthier, healthier and able to access far more services and opportunities than the working class (see Box 5.1). Many argue, as Robison (1997) notes, that the middle

Box 5.1

Choosing birthdays: wealth, power and the middle-class life in Thailand

The urban middle classes of Thailand are becoming important actors in Thai development. Their wealth has allowed them to adopt habits and interests that mimic those of a globalised middle class. They shop in sprawling mega-malls that offer the latest and trendiest consumer goods from around the world; they watch Hollywood films in massive multiplexes (see Figure 5.1); they are gentrifying inner city neighbourhoods by moving into new modern comfortable apartment blocks; or are seeking out new kit homes in the suburbs of estates with Western sounding names such as Hampton Residence, Madison, Chateau de Bangkok, Windsor and British Town. At the same time

Figure 5.1 ***Middle-class mega-mall shopping in Bangkok.***

Source: Author

however the middle class retains its distinctly Thai identity by continuing to frequent Thai-style malls in which tiny stalls sell everything from pirated CDs and imitation Gucci bags to mobile phones and *pad thai* foodstalls; they still hail down three-wheeled *tuk-tuks* to transport them around town; and many continue to prefer a tasty bowl of fried grasshoppers to a serving of lasagne or steak.

The wealth of the middle class opens up new opportunities that are unavailable to lower income groups. In the north of Thailand, for example, urban middle-class women in Chiang Mai have a much greater array of birthing options available to them than low income rural women. As Sinjai, an urban middle-class woman, remarked:

> When I had my first child, [X] hospital was the best hospital in town. And when I had my second child, [Y] hospital was the best, as it has everything and it was very convenient and comfortable for me. Even the room was very good. My doctor worked there as well.
>
> (Liamputtong 2005: 253)

In contrast, Payao, a working-class rural woman who was lucky enough to secure funds to travel to Chiang Mai, told of her fears of visiting her local hospital:

> I went to the maternity hospital in Chiang Mai city as people told me that there would not be too many diseases in that hospital as it is only for the mothers and babies. If I went to our local hospital or other public hospitals, there would be a lot of diseases around as there are many patients with different diseases and as we have to mix with them, the chance of getting some diseases would be high.
>
> (Liamputtong 2005: 253)

While antenatal care is free in public hospitals middle-class women can choose to have their own private obstetricians, which can cost 17,000 baht (US$472) for a vaginal birth to 30,000 baht (US$833) for a caesarean. While the option of caesarean births may reflect globalisation, the reasons behind them, as was the case for Wilai, may be distinctly Thai:

> My husband wanted the baby to be born on either a Monday or Thursday, but not a Saturday. He said a Monday child will be clever and smart and if a baby is a boy he will become a soldier. A Thursday child will be a strong child and easy to raise. A Saturday child will have a dominant fate which means he or she will be stubborn and may become aggressive and parents will have problems in bringing them up. The baby was due around Thursday and because my doctor's schedule was not too busy Thursday, we chose a Thursday.
>
> (Liamputtong 2005: 257–258)

The freedom of Wilai to choose a Thursday child as well as how and where the child will be delivered, emphasises the power middle-class populations are

are gaining over their bodies and lives. It also emphasises the gap, which begins at before birth, between middle-class lifestyles and those of the rural poor.

Sources and further reading: Cornwel-Smith (2005), Liamputtong (2005)

classes are new liberalising elements in society who will protect their interests by breaking down authoritarian governance structures and constraints on individual liberties and freedoms. In this line of argument they are portrayed as agents of political and social change who will reform repressive elements of society through their actions. This can be seen in the proliferation of middle-class NGOs and student unions in parts of the region, focusing upon things such as human rights, governance, corruption, environmental degradation, education, health, land rights, poverty and a range of other issues that cross class boundaries. On the other hand they can also be seen as a conservative group who progressed to this point through hard work and thrift, the types of attributes associated with the Asian values argument discussed in Chapter 4. In this view the middle classes are self-interested and display common capitalist middle-class traits of political apathy and conspicuous consumerism and are much more likely to use existing structures, repressive or otherwise, for their own personal ends rather than to revolutionise society. Mulder (1997: 111) refers to this when critiquing some younger middle-class Filipinos: 'Most members of the new urban middle class are upwardly mobile, mass-educated, and directed more to professional advancement than to ideological reflection . . . They grew up with television rather than books.' Both perspectives are likely to be correct as the sheer number of people now in the middle class makes it more appropriate to refer to the middle class*es* rather than consider them as a single homogeneous unit. Hence while liberal middle-class university students have been important agents of political change in countries such as the Philippines, Indonesia, Thailand and even Myanmar, it is likely that other middle-class groups would have been horrified by their actions given the potential threat such change could pose to their lifestyles.

At the top of the economic chain in Southeast Asia are relatively small groups of wealthy elites. The elite class is comprised of the families of successful businessmen (and they usually are male), politicians and military figures who have managed to capture a significant slice of the profits arising from rapid economic development. In most Southeast Asian countries the state and military elites are usually of indigenous origin whereas business elites are dominated by those of Chinese ethnicity (see Chernov 2003). It is common for elites from commercial, state and military sectors to all mix in the same circles and even shift between jobs and sectors, from the head of a powerful corporation to a political appointment for example, to look after one another's interests and ensure harmony at the highest levels (Case 2003). While elite groups have always existed in Southeast Asian societies rapid economic development has greatly enhanced their power, wealth and influence. Lim and Stern (2002), for example, write that the wages of Singaporean bureaucrats are matched to those of US CEOs and the Singaporean prime minister earns several times that of the US president. Elites generally have an interest in repressing civil society as social movements, NGOs and labour unions are likely to challenge their disproportionate share of wealth and power.

The stratification of Southeast Asian societies into different classes as a result of economic development creates new tensions and alliances for social and political development. If class interests, particularly middle-class interests, are not protected civil society activism is likely to increase and pressure will mount for social and political change. In Singapore, Malaysia and Brunei middle-class interests appear satisfied with the share of wealth and resources that they capture from their national economies. This has led to a tacit alliance between middle and elite classes that has contributed to the suppression of labour unions and working-class interests in these countries. In Thailand, Indonesia and the Philippines, however, temporary alliances have formed between middle and working classes, and sometimes disaffected elites, which have led to widespread protests, strikes and political pressure. In Indonesia, for example, the Asian crisis brought hardship to most and led to an alliance between middle and working classes who mobilised through protests and strikes to bring about President Suharto's resignation and the subsequent restructuring of Indonesian society and politics. Similarly in the Philippines President Marcos ostracised workers and the middle class, as well as many elite families by creating a new generation of 'crony capitalists' made up of relatives and friends whose power threatened the wealth of traditional business elites. A temporary alliance between these classes and the disaffected elites saw thousands of people take to the streets in the People Power Revolution that lead to Marcos' fall from power (see Case 2003). With Laos, Vietnam and Myanmar now attracting more FDI and Cambodia and Timor-Leste following a similar path to the ASEAN-5 states it is likely that development will similarly favour some more than others leading to similar class divisions. To secure equitable forms of development it is important that civil society is free to organise so collective interests and concerns can be identified, voiced and resolved. It is also important to remember that the majority of Southeast Asian people still live in increasingly stratified rural areas and that their interests should be similarly recognised and resolved through civil society networks and processes (see Chapter 7).

Religion and development

Southeast Asia possesses diverse spiritual landscapes. In more remote areas localised indigenous animist belief systems remain influential. Animists see local spiritual forces and entities, such as ancestors or creation spirits, in the trees, rocks and objects of the physical landscape. More influential however are the religions of Islam, Buddhism and Catholicism, which as explained in Chapter 2, were transported to the region by traders, missionaries and colonialists. Islam is the most popular religion being practised by around 210 million people across the region, most of whom are located in Indonesia, Malaysia, Brunei and southern parts of the Philippines and Thailand (Fealy 2004). Buddhism dominates mainland Southeast Asia while Catholicism dominates most of the Philippines and Timor-Leste. Confucianism shapes Buddhist interpretations in Singapore and Vietnam, while Hinduism is practised on the Indonesian island of Bali. In reality all countries harbour a range of religious beliefs and should be considered plural religious societies.

Colonialism and development have gradually reduced the power and influence of religious institutions. All the mainstream development theories described in Chapter 1 prioritised economic and social development above spiritual concerns and implemented secular development projects. People now look to the state for education, health and governance, whereas religious institutions once played an important role in these sectors in the past. Religious responsibility is being reduced to purely spiritual matters, representing a significant downsizing of their pre-colonial power and influence. Nevertheless within these spiritual domains religion is still very important and shapes people's everyday behaviour and provides them with a lens to interpret development processes. On the island of Bali, for example, people still find time to make small offerings or *banten* (see Figure 5.2) on a daily basis and retain important religious ceremonies despite the rapid tourist-driven changes taking place around them. Similarly Muslims in Indonesia and Malaysia are still called to prayer five times

Figure 5.2 Balinese offering (banten).

Source: Sarah Ellis

a day and Islamic norms govern issues of appropriate dress and behaviour. In Buddhist countries many boys are still expected to serve time as monks when they learn Buddhist principles that do not necessarily accord with the norms of modern development. The importance of religion to people's everyday lives can be seen in the impressive religious structures that abound; it seems even the poorest and most marginalised communities are able to find funds to materialise their beliefs (see Figures 5.3–5.5).

While development may have lessened the authority of religious institutions in social, economic and political matters, they retain an important role in civil society. Religious institutions have incredible networks that link remote rural communities with their urban counterparts and they have retained considerable respect and authority at the community scale. It is within religious spaces that the merits, morals and problems associated with development are often discussed. In mosques across Indonesia and Malaysia, for example, people have a place to discuss their moral concerns about development and, if necessary, organise formal or informal protests to lobby for more sensitive or correct approaches. In the Philippines churches have played

Figure 5.3 Rahmatullah Mosque, Lampuuk, Aceh, being rebuilt after 2004 tsunami.

Source: Author

an active libratory role by supporting the People Power Revolution and providing a space for peasants to organise and lobby for land reform (Carroll 2004). In Thailand monasteries have become active in educating people about more sustainable Buddhist approaches to nature, and have established forest monasteries where land is protected from logging and clearance (see Chapter 8).

These types of activities have led to comprehensive religious statements being expressed about the nature and focus of development. This is most pronounced in Thailand with King Bhumibol's vision of the 'sufficiency economy' (see UNDP 2007). The sufficiency economy draws on principles of Buddhist economics to argue for a uniquely Thai way of developing that opposes many of the fundamental principles of mainstream economic development theories. The king's most complete statement came on his 1997 birthday address to the nation when it was reeling from the Asian economic crisis. During his address he argued for 'a careful step backwards' to ensure a 'self-sufficient economy' through three stages. The first stage advocated self-reliance at the household scale; the second stage extended self-reliance to the community by mobilising the surplus resources of households

Figure 5.4 Chedi at Wat Phra Chetuphon, Bangkok.

Source: Kara Barnett

and forming cooperative forms of production; once communities were secure they could engage with trade beyond the village to gain new technologies and access to financial institutions for mutual advantage, the third stage. A Thai working group has developed the basic principles of the sufficiency economy as follows:

- *Moderation*: a middle way between want and extravagance, between backwardness and impossible dreams.
- *Reasonableness*: evaluating the reasons and short- and long-term consequences of any actions on oneself, on others and upon nature.
- *Self-immunity*: the ability to withstand shocks and unpredictable events, to be self-reliant.
- *Knowledge*: Accumulating wisdom and putting it to use in prudent and useful ways.

Figure 5.5 Church in Dili, Timor-Leste.

Source: Author

- *Integrity*: Acting with virtue, honesty and ethically; working hard and refusing to exploit others.

At the core of the sufficiency economy are Buddhist principles that are in stark contrast to the core foundations of mainstream economic development theory. Conventional economics, for example, argues that people are self-interested and they will gain happiness by maximising their capacity to consume, through the generation of economic wealth. Buddhists, on the other hand, distinguish between two different types of desire, one is materialistic and focused on consumption, while the other is focused on quality of life. According to Buddhist economics the desire to consume is a false desire and efforts should instead be directed towards achieving quality of life rather than maximising consumption (Prayukvong 2005; see also Schumacher 1993). The desire to maximise consumption has no limits and as such it breeds unhappiness, anxiety, competition, degradation and conflict in a world of finite limits. In becoming self-sufficient and focusing on the community, people can become content, improve their quality of life and develop in equitable, sustainable and harmonious ways. The sufficiency economy has found a receptive audience in some places and a variety of community scale development initiatives build upon its basic foundations (see UNDP 2007; Prayukvong 2005; Suwanbubbha 2003). It is less acceptable to entrenched urban elite and middle-class interests, as they are currently benefiting from the uneven patterns of conventional economic development.

Islamic principles are increasingly influencing development in the Muslim states of Malaysia, Indonesia and Brunei, and also in sub-national spaces such as the Philippines autonomous region of Muslim Mindanao (see Chapter 4). Islam is becoming increasingly popular and consequently politicised within Malaysia as the ruling UNMO and the main opposition party, the Pan-Malaysia Islamic Party (PAS), compete prove their religious credentials to their Islamic populations. Under Prime Minster Mahathir, and later under his successor Abdullah Ahmed Badawi, UNMO has championed the concept of 'Islamic modernisation' to the rest of the world, referring to a liberal interpretation of the Qur'ān and the application of Islamic principles to economic and spiritual development. The country declared itself an Islamic state in September 2001 and Islamic teachings have increasingly influenced state services such as courts, schools, media and research (Hamayotsu 2002; Martinez 2002; Fealy 2004). The sizable proportions of non-Muslims in the country have offered little resistence due to the progressive nature of UNMO's interpretations. For PAS, however, UNMO is too liberal in its reading of the Qur'ān and they have lobbied for a more conservative Islamic society based on the application of Islamic laws such as the shariah (see Kadir 2004). In some areas local Islamic courts have been established that rule on shariah issues such as theft, adultery, drinking and dress. While the shariah is not popular with everyone it does boost the Islamic credentials of PAS and is putting pressure upon UNMO to adopt a more Islamic development pathway.

Religious civil society movements in Indonesia are also beginning to lobby for Islam to play a bigger part in national development plans. Under General Suharto, Islamic parties were often portrayed as threats to the unity of the state, going against the principles of religious tolerance outlined in the national development ideology of *pancasila* (see Chapter 4). Islamic movements also potentially threatened the position of wealthy Chinese elites, most of whom are non-Muslim (Hamayotsu 2002). Late in Suharto's career, however, he became more influenced by Islam and allowed for the formation of the Indonesian Association of Muslim Intellectuals that shaped much of his own, and his successor, President Habibe's, policies. The post-Suharto shift to democracy has seen Muslim groups become increasingly active and influential in civil society and political circles, with subsequent presidents having strong links to various Islamic institutions (Baswedan 2004). While few are pushing for a strict Islamic state, like that seen in Iran, shariah courts have been established in various areas and, as is the case in Malaysia, some aspects of Islamic economics have been trialled. Islamic economics refers to the prohibition of *riba*, which is interest applied to capital, and the application of *zakat*, which is the provision of alms to the poor. Instead of charging interest Islamic banks lend money in return for a proportion of the profits, but receive nothing if the business fails. This makes them less likely to provide money for risky investments and prevents people from going into debt if their business fails. Under *zakat* Muslims are required to redistribute some of their wealth to the poor, the handicapped, needy travellers or teachers of Islam for purification and positive social transformation. While both these approaches face considerable challenges in their application they do suggest an Islamisation of conventional development theories is taking place opening up new ways of pursuing more equitable patterns of development (see Hefner 2006 for a discussion).

Gender and development

The status of women in Southeast Asia is often compared favourably with many other parts of the world. Women are active in private and public life, are productive members of society and are highly valued across the region. The Philippines, with Presidents Corazon Aquino (1986–1992) and Gloria Macapagal-Arroyo (2001 onwards), is one of only a handful of countries that can boast more than one female head of state while Indonesia elected Sukarno's daughter, Megawati Sukarnoputri, as president from 2001–2004. The head of main opposition party in Myanmar, Aung San Suu Kyi, has become a figurehead for democracy movements around the world, having been awarded the Nobel Peace Prize for her non-violent resistance to military rule. Most countries now have women's representative bodies at the national scale that bring gender considerations into everyday planning and women are found in most occupations earning a range of salaries (see Kusakabe 2005). However, like women the world over, they also suffer discrimination that sees them generally paid less, placed in less prestigious occupations and unrecognised for their household labour responsibilities. Many suffer discrimination in terms of education, health care and responsibilities throughout their lives.

One useful way to think about the multiple roles of women within Southeast Asian development is to think in terms of their differing reproductive and productive roles within society (for a fuller discussion see Momsen 2004). Reproduction refers to both the biological and social roles that are acted out to enable the conditions for families, households and societies to be maintained. Hence biological reproduction refers to childbearing and early nurturing of infants whereas social reproduction refers to household labour such as cleaning, cooking, collecting water and firewood and caring for the sick. Productive labour, on the other hand, refers to activities that have some sort of exchange value, usually cash income. Traditionally in Southeast Asia, as in most parts of the world, women have far more reproductive responsibilities than men, while men have tended to dominate productive roles (although women are also heavily involved). While social development, in terms of the expansion of health services, schools, water supplies and marketplaces may have assisted with women's reproductive roles, there has also been increasing pressure for women to partake in new productive opportunities associated with development (see Eder 2006; Kusakabe 2004 for examples). The advantage of this is that women become less dependent upon their partners for financial support and there are more opportunities in their lives. The disadvantage, however, is that in taking on new productive roles there is often minimal reduction in their traditional reproductive roles, effectively adding new responsibilities and duties to already busy lives.

Some of the new productive opportunities for women include work in service industries such as sales, tourism and hospitality, as well as factory work in the textile, footwear and electronics industries. For young rural women these opportunities often require migrating to the city for a few years to earn money before returning home to undertake more conventional household reproductive responsibilities. Many educated or skilled women, particularly from the Philippines, seek employment overseas where they can earn higher incomes as nannies, nurses or domestic help (see Box 5.2). In most cases migrant women

Box 5.2

Flor Contemplacion and the Filipino export labour industry

In 2006, approximately 5 million Filipinos were working overseas, almost 10 per cent of the domestic population. Approximately 650,000 move every year, mostly for economic reasons as potential incomes can be up to twenty times higher than the income earned for equivalent work at home. Most migrants are temporary and retain important links to their Filipino households, and the broader Filipino economy, by sending portions of their salaries home as remittances that support household costs. The combined value of these remittances reaches US$8–10 billion annually for the Filipino economy, comprising approximately 18 per cent of the Philippines gross national product (GNP) over the past decade. Men initially dominated this industry, finding jobs in the construction sector of booming Middle Eastern oil economies in the 1970s, however economic migration is now dominated by women who find jobs in Asia, North America, Europe, the Middle East and Australasia as domestic helpers such as maids or nannies; health care workers such as nurses; and 'entertainers', which can include dancers, hospitality workers and sex workers. The government encourages and facilitates the export of Filipino labour as a national development strategy, providing official receptions and awards for its overseas working 'heroes'.

One of the most famous and tragic overseas 'heroes' is Flor Contemplacion. For many she has become a symbol for the risks involved in overseas economic migration, particularly for domestic workers. Contemplacion was one of approximately 100,000 overseas workers, 75 per cent of whom were Filipino, who work as maids in Singapore. In 1991 Contemplacion was accused by Singaporean authorities of murdering her friend Maga, another domestic worker, as well as Nicolas Huang, the young boy Maga was looking after, during a fit of insanity and rage. Subsequent investigations found a number of discrepancies in the Singaporean case and most Filipinos believe Maga's employer, Wong Sing Keong, killed Maga after discovering Nicolas had died in the bath while having an epileptic fit. Singaporean officials never informed Contemplacion's family that she had been arrested and they only found out in 1995 after the Filipino media had escalated the case into an international incident. Her children, who were her original financial motivation for leaving the Philippines, flew to Singapore as the Filipino government downgraded diplomatic relations in protest. They were not allowed to embrace her before she was hanged for murder in 1995.

Contemplacion's case emphasises the difficult positions domestic workers find themselves in:

> The foreign domestic worker as a foreigner and a domestic occupies an untenable space – neither incorporated as an employee in the public

> sphere with social and legal rights under the jurisdiction of the state, not a member of the familial where relations are governed by non-market affinities.
>
> (Yeoh and Huang 1998: 588)

Domestic workers have few legal rights, supportive mechanisms or means to protest abusive working conditions. If conflict arises they risk losing their visas and authorities are much more likely to believe employers' versions of events over foreign employees. Migrant workers may experience new freedoms, opportunities and financial power by travelling overseas, however it can come at great personal cost and sacrifice; in the case of Contemplacion it was much more than anyone could be expected to give.

Sources and further reading: Ofreno and Samonte (2005), Meerman (2001), Hilsdon (2003), Yeoh and Huang (1998)

send the majority of their earnings home and their relatives, friends or occasionally hired help cover their usual reproductive roles. The move into paid employment opens up new experiences and opportunities for such women although it also exposes them to new risks and vulnerabilities. While many women choose to participate in new development opportunities others are forced into productive roles due to financial hardship. This was particularly the case during the Asian financial crisis when many men lost their jobs and inflation was pushing up the cost of basic foodstuffs and household goods. To counter these effects many women took on additional productive responsibilities and experimented with informal small businesses activities and low paid employment. On the island of Sulawesi in Indonesia, for example, King and Kim (2005) found that the financial crisis resulted in women instigating local credit associations and becoming increasingly involved in non-agricultural income generating activities such as petty trading, selling snacks and managing small shops or *warungs*. These income-generating activities effectively acted as 'socio-economic buffers' to the negative impacts of the economic crisis. While this boosted household income it was rarely accompanied by a reduction in reproductive duties that were seen as women's work, even if unemployed men had more time on their hands. In these cases development has lead to a sharp increase in women's duties and workloads.

Women's movements and NGOs have sprung up across the region to represent their interests within development. Many established important international networks during the United Nations Decade for Women (1976–1985) and have since found support from international donors keen to improve the status of women within development. Women's movements have focused attention upon the gendered impacts of development, highlighting the economic focus of most mainstream development programmes and the ramifications this has for women in both productive and reproductive roles. Some NGOs have sought gender equity in income, others have provided services and advice, and others have sought to educate women on their basic human rights. Women's movements are particularly active in the Philippines where Sobritchea (2002) has identified the following sectors as their key foci:

- *Welfare services* – home visitation, shelter for abused women and children, legal services.
- *Gender training and education* – gender awareness, leadership, reproductive health and rights, informal education.
- *Health and medical services* – crisis counselling, reproductive health and general women's health issues.
- *Livelihood and small business* – microcredit and small business development.
- *Advocacy* – gender and development, violence against women and children, sex trafficking and prostitution, sexism in the media and economic reforms.
- *Research* – on all the topics listed above.

Outside of these general areas women's movements have also sought to expose patriarchy in religion. Women's NGOs in Thailand, for example, have sought to re-establish the ordination of the female religious order, the *bhikkhuni sangha*, of Theravada Buddhism in Thailand (Tomalin 2006), while Sisters of Islam has worked closely with UNMO to update and reform patriarchal interpretations of shariah law in Malaysia (Yan 2002). While women's NGOs do not have the same freedom to form and lobby for change within non-democratic countries women's interests are still represented by GONGOs such as the Women's Unions of Laos and Vietnam as well as the Myanmar Maternal and Child Welfare Association.

One sector that women's NGOs are particularly concerned about is the sex industry that is extensive in parts of Southeast Asia, particularly within Thailand and the Philippines. Views on sex workers differ greatly; some see them as unfortunate victims of globalisation and development while others see them as empowered workers taking advantage of a lucrative industry. The most vulnerable women are those that have been forced into prostitution through desperation or trafficking (see Box 5.3). Beyrer (2001), for example, has found that a disproportionate amount of sex workers in Thailand are from the Shan ethnic group in Myanmar, being forced from state persecution at home to a more personal form of exploitation in Thailand. Similarly Taylor (2005) has explored the causes of trafficking of rural girls in the north of Thailand to the sex centres further south, pointing out that not only poverty but also a lack of opportunity encourages families to send their daughters away. Many of those 'sold' into the sex industry have very little say in the decisions made for them and must pay back the brothel owner before they can receive any financial return for the services they are providing. Such exploitation is an adverse impact of development and globalisation and many NGOs are now operating in Southeast Asia to protect sex worker rights and health, particularly in relation to sexually transmitted diseases such as HIV/AIDS (see Marten 2005; Law 1998). Women's NGOs also seek to provide services and training that can assist women in finding alternative employment should they chose to leave the sex industry. These types of activities are vital in combating exploitation and expanding the range of opportunities available to women in spaces of rapid economic development.

Box 5.3

Human trafficking in Southeast Asia

> When Siri wakes it is about noon. In the instant of waking she knows exactly who and what she has become. As she explained to me, the soreness in her genitals reminds her of the fifteen men she had sex with the night before. Siri is fifteen years old. Sold by her parents a year ago, her resistance and her desire to escape the brothel are breaking down and acceptance and resignation are taking their place.
>
> (Bales 1999: 34)

Stories of the human degradation of trafficked persons, particularly women and girls, have garnered increased attention by journalists, academics, interest groups and the development community around the globe. The majority of the stories told are similar to the one related above, recounting the horrors of women and children trafficked into prostitution. Although an important aspect, the trafficking of women and girls into prostitution represents only one dimension of the larger process.

According to article 3 of the Trafficking Protocol (United Nations 2000: 32), which supplements the United Nations Convention against Transnational Organised Crime:

> 'Trafficking in persons' shall mean the recruitment, transportation, transfer, harbouring or receipt of persons, by means of the threat or use of force or other forms of coercion, of abduction, of fraud, of deception, of the abuse of power or of a position of vulnerability or of the giving or receiving of payments or benefits to achieve the consent of a person having control over another person, for the purpose of exploitation. Exploitation shall include, at a minimum, the exploitation of the prostitution of others or other forms of sexual exploitation, forced labour or services, slavery or practices similar to slavery, servitude or the removal of organs.

Deceived, lured, and sometimes even kidnapped or sold, victims are forced into situations where the conditions are often indistinguishable from slavery. Southeast Asia is renowned for human trafficking. The trafficking of men, women and children occurs amid high levels of internal and cross-border migration, and it occurs in a wide range of settings and for a variety of purposes such as begging, domestic help, marriage, labour in factories, work on fishing boats, and prostitution. In search of better opportunities, most lose basic human rights, dignity and control over their lives through threats of violence, coercion and debt-bondage.

Human trafficking trends in the Southeast Asian region take many forms. For example, in Cambodia, both women and men are trafficked internally as well as into neighbouring countries such as Thailand for the purposes of commercial sexual exploitation, and, more recently, into the construction, agricultural and fishing industries. Children are exploited as well. Taken from familiar surroundings, they are forced to beg on the streets of Thailand and Vietnam. Those living in poverty stricken countries with high proportions of young people, such as Laos and Myanmar, are also particularly vulnerable to trafficking. Most end up in Thailand where they are undergo forced labour and commercial sexual exploitation. Thailand is a key regional destination country and is also the major transit and sending country in the region. With its economic pull unlikely to diminish soon, Thailand must grapple not only with human trafficking but also with the flow of illegal migrants. Increasing demands for cheap, exploitable labour, as well as limited legal migration channels, leaves migrants more vulnerable to traffickers.

It is not surprising that human trafficking has become one of the most complex and pressing global challenges facing the development community. Progress has been made by scholars and practitioners in improving our understanding of human trafficking, targeting research and developing effective counter trafficking measures. Yet, the demand and supply for trafficked persons is endless. The reality is that we have only begun to comprehend the enormous scope of the problem.

Rebecca Miller, Centre for Development Studies, University of Auckland

Sources and further reading: Bales (1999)

Conclusions

Social movements and civil society organisations are very much part of development within Southeast Asia. Some are mobilising in response to their new identities that development has thrust upon them while others are trying to imprint their more traditional identities onto development. Civil societies give a Southeast Asian flavour to development processes and show that ordinary people, as much as economic and political systems, can be important forces of social change. An active civil society is important from an equitable development point of view as it allows undesirable development processes to be recognised and rectified and gives people a chance to influence development in ways that they consider most appropriate. Whether this results in a greater share of wealth or improved working conditions for working-class people through the trade union movement; or the adoption of religious development policies based on Buddhist, Catholic or Islamic principles; or greater recognition and support for the disproportionately high productive and reproductive workloads of women; or a host of other social issues not covered in this chapter, civil society and social spaces are necessary to create appropriate, fair and people-centred development. Current restrictions on civil society processes are contributing to the uneven development patterns found across Southeast Asia.

Summary

- **Civil society movements are expanding in some parts of Southeast Asia and are empowering people to become important actors within development.**
- **In other places the state restrains autonomous civil society organising and dictates the types of issues they are allowed to pursue.**
- **New class-based institutions, particularly middle-class organisations, have the potential to transform or entrench current approaches to civil society.**
- **Religious and women's institutions are transforming development at local and national scales.**
- **From the perspective of equitable development a free and active civil society is vital to the recognition and protection of marginalised communities.**

Discussion questions

1. How are civil society institutions shaping Southeast Asian development?
2. Discuss the impact of development upon women's lives in Southeast Asia.
3. Discuss the strengths and weaknesses of trade unions from the perspective of different social classes.
4. What are the ramifications of religious movements for Southeast Asian development?

Further reading

Two insightful edited collections with case studies of Southeast Asian civil society are:

Guan, L. (ed.) (2004) *Civil Society in Southeast Asia*. Singapore: ISEAS Publications.

Weller, R. (ed.) (2005) *Civil Life, Globalisation, and Political Change in Asia: Organising between the Family and the State*. London: Routledge.

An introduction to religious approaches to development can be found in both:

Hefner, R. (2006) Islamic economics and global capitalism. *Society* 44(1): 16–22.

UNDP (2007) *Thailand Human Development Report 2007: Sufficiency Economy and Human Development*. Bangkok: UNDP.

Good texts on gender issues and development include:

Momsen, J. (2004) *Gender and Development*. London: Routledge.

Bahramitash, R. (2005) *Liberation from Liberalisation: Gender and Globalisation in Southeast Asia*. London: Zed Books.

Useful websites

The most important institution looking into labour rights issues is the International Labour Organisation, which has a useful website at: www.ilo.org

6 Transforming urban spaces

Introduction

The economic, political and social changes discussed in earlier chapters manifest themselves differently in urban, rural and natural spaces. This chapter concerns itself with urban spaces in which over 50 per cent of Southeast Asia's population are expected to live by 2017 (Jones 1997). While the current proportion of people living in urban centres is still low compared to the Americas, Europe and Oceania, the popular rural image of Southeast Asia as a region of paddy farms and village lifestyles is becoming less common as more and more people migrate to the cities. Today Southeast Asia is a region of bustling metropolises, high-rise skyscrapers and burgeoning urban sprawl. In many ways the expanding metropolises of Southeast Asia have come to symbolise the region's successful pursuit of development, signifying the emergence of the region within global economic networks and flows. Cities and their hinterlands have attracted the bulk of FDI and, in most cases, are far more developed in terms of economies and social services than rural spaces. However one does not have to scratch far beneath the urban mosaic to realise that with development successes comes new challenges that planners will continue to grapple with for years to come.

Development and urban typologies

In the fifteenth to seventeenth centuries Southeast Asia had cities with populations approaching 100,000 people making them larger than most European cities of the time (Reid 1980). McGee (1967) has classified these early cities into sacred and market cities. Sacred cities were those powerful centres that housed 'godkings', rulers, clergy, senior military figures and associated administrative processes. Market cities, on the other hand, grew from their position on international trade routes, becoming important centres of exchange and commerce that housed traders from around the world (see Forbes 1996 for a discussion). While these ancient cities form the foundations of many of the region's metropolises today other contemporary cities trace their roots to the colonial period. Singapore, for example, had a role in the Srivijaya Empire but had shrunk to little more than a fishing village before the arrival of Sir Stamford Raffles who founded the city in 1819. Similarly, Ho Chi Minh City (formerly Saigon), in the south of Vietnam, and Manila in the Philippines, only emerged as important cities after considerable French and Spanish colonial investment.

Since independence Southeast Asian cities have continued to grow and diversify. One urban form that is common for all states in Southeast Asia is that of the capital city. Capital cities are urban areas that house the centre of government and administration and are consequently places of power where development policy and decision-making are undertaken. Economic development, service industries, transport infrastructure and incomes are usually concentrated in capital cities and their hinterlands as it is here that TNCs and aid industries set up their offices so they can be in frequent contact with power brokers. Capital cities form the symbolic heart of many Southeast Asian countries and are important in nation-building activities oriented at developing national identities (see Chapter 4). They are often littered with museums, galleries and important monuments that signify national achievements aimed at inspiring pride and unity within the nation. In Kuala Lumpur, for example, when the Petronas Twin Towers were built in 1996 it was the tallest building in the world; proudly symbolising Malaysia's technical development success (see Figure 6.1; Bunnell 1999). Bangkok has the historic Grand Palace that houses Wat Phra Kaew and the Emerald Buddha, as well as Wat Phra Chetuphon's gold-plated reclining Buddha (see Figure 6.2), while

Figure 6.1 Petronas Twin Towers, Malaysia.
Source: Jane Dunlop

Yangon[1] has the huge Shwedagon Pagoda, all of which play important symbolic roles within national identities. Hanoi emphasises national unity though the Ho Chi Minh mausoleum where people pay their respects to the most respected figure in Vietnam's independence movement. Nation-building strategies can also bind people together through memories of shared pain, such as Phnom Penh's Tuol Sleng Museum of Genocide Crimes that delves into Cambodia's difficult past (see Hughes 2003). Capital cities also represent the 'face' of a country to outsiders with more impressive services and infrastructure

[1] The official capital of Myanmar was shifted from Yangon to the relatively unknown and inaccessible location of Pyinmana in 2005 (see Jagan 2006 for a discussion of possible reasons). For economic and social purposes however, if not political and military, Yangon remains the centre of activity.

Figure 6.2 Reclining Buddha in Wat Phra Chetuphon, Bangkok.

Source: Kara Barnett

likely to attract more investment, tourism and skilled international migration.

Another type of city located within Southeast Asia are world cities. These are cities that play important roles within global economies, being centres for international finance, banking and telecommunications. In a study by Beaverstock *et al.* (1999) Singapore was ranked as one of the globe's ten most important and influential world cities. This is an impressive achievement, particularly when considering that Tokyo and Hong Kong were the only other top ranked world cities outside of Europe and the US. Singapore's presence in this list reflects its colonial history as an important port for East–West trade, as well as its early adoption of EOI principles that successfully attracted large volumes of FDI (see Chapter 3). This outward looking development strategy has resulted in Singapore housing the region's most influential stock exchange, the most important banking, financial and service centres, as well as the largest number of regional headquarters for TNCs (see Smith 2004). It has attracted a skilled educated global workforce through tax breaks, high wages, fast transport, high-tech infrastructure and an ongoing city beautification programme that is making it a more liveable city for expatriates. Singapore's high profile and positive urban development ensures it attracts a steady stream of tourists as well as businesses wishing to invest or expand into the region. According to Beaverstock *et al.* (1999) Bangkok, Kuala Lumpur, Jakarta and Manila also qualify as world cities, albeit with substantially less power and influence than Singapore. Each of these cities is a capital city and consequently a centre of domestic economic and political power, they all sided with the West during the Cold War and they all experimented with EOI strategies relatively early in the post-colonial era. This combination of factors has made them important regional hubs for foreign investment and production.

Primate cities dominate national economies and contain a disproportionately large percentage of a country's urban population. Primate cities in Southeast Asia include almost all the capital cities of the region with the exception of Hanoi, where Ho Chi Minh city and Hai Phong are also of significant size, but is particularly evident in Manila and Jakarta, which at over 10 million people are considered mega-cities, and Bangkok which, in the early 1990s, was estimated to be forty times the size of the next urban settlement (Askew 2002: 59). Primate cities are most pronounced in the uneven market-led economies of Southeast Asia where service and manufacturing industries as well as transport and communications infrastructure has been concentrated into single locations. This attracts FDI as well as workers from other regions seeking higher wages and better conditions (see Figure 6.3). As primate cities grow, however, they become less economically efficient as the cost of land rises, transport times increase and health deteriorates with rising pollution. In Bangkok, for example, there are 2 million cars (approximately 50 per cent of the country's total), which average 1–2km/h at peak hour, contributing to carbon monoxide levels that are often fifty times World Health Organisation standards (Glassman and Sneddon 2003). These negative impacts are enhanced when population growth is rapid and unchecked as it strains the city's abilities to provide housing, water, waste collection, education and employment to the rapidly expanding population

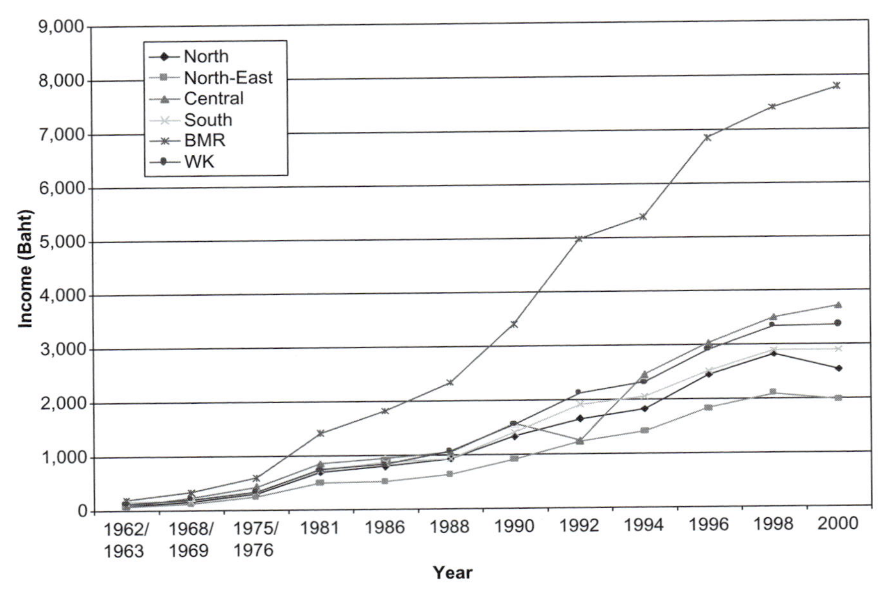

Figure 6.3 Urban primacy and monthly incomes in Thailand.

Source: Krongkaew and Kakwani (2003)

Note: BMR – Bangkok Metropolitan Region; WK – Whole Kingdom.

(Henderson 2002). To counter these emerging problems national authorities tend to invest more money into the primate city to alleviate transport and pollution issues, which then detracts from the money available to boost urban development in other centres. Hence urban primacy can detract from economic development elsewhere because of the costs associated with building and maintaining a core economic centre. Governments in Southeast Asia are recognising the inequalities, constraints and unevenness associated with primate cities and are experimenting with policies to grow alternative economic centres.

Related to primate cities in terms of their economic dominance are extended mega-urban regions (EMRs). This term refers to cities that have expanded outwards from their original core to incorporate neighbouring settlements that were once separate entities. The large primate cities of Manila, Bangkok and Jakarta, as well as Kuala Lumpur and Singapore, through the IMS-GT (see Box 3.2), are all examples of EMRs (McGee 1995). EMRs generally have a *city core* that is based on the original city limits; a surrounding *metropolitan area* that bounds the heavily built up area of the core; and an *extended metropolitan area* that radiates outwards along transport routes to incorporate surrounding areas. Hence Bangkok has the original

Bangkok Metropolis, the Vicinity and the Extended Urban Region, while and the original core settlement of Jakarta is surrounded by the Jakarta Metropolitan Area (or Jabotabek) that incorporates the nearby municipalities of Bogor, Tangerang and Bekasi, before reaching into the extended metropolitan region along arterial transport routes. Different economic activities take place in each of these zones with the core areas dominated by tertiary industries such as finance, sales and hospitality; the metropolitan areas are dominated by industrial manufacturing activities; and the extended metropolitan area is typified by a mix of industrial activities and rural agriculture. Indeed many places in this outer zone retain much of their village character having only been incorporated into the EMR by their chance position between two previously disconnected urban centres. McGee (1991) has referred to these areas as *desakota*, which roughly translates to village-city, where rural-style villages exist within a broader urban setting. Such villages now play an important role in growing agriculture produce for urban consumption (see Box 6.1), and also

Box 6.1

Farming the city: urban agriculture and *doi moi* in Hanoi

Urban agriculture plays an important role in sustaining urban livelihoods and supplementing the subsistence food needs of urban communities in Southeast Asia. There is a range of reasons why agriculture is pursued in urban areas. Some farmers have never given up their livelihoods despite urbanisation processes taking place around them, other farms have recently been swallowed into *desakota* cities as a result of urban sprawl, and newer agricultural enterprises have been initiated by rural migrants applying the skills they know best. Consequently it is not unusual to see vegetable fields planted beside high rise buildings, or livestock grazing quietly beside a busy road.

In Vietnam's capital, Hanoi, agriculture is a well-established and highly organised practice that extends back to the original formation of the city. Since 1986, however, it has undergone some major transformations as a result of the *doi moi* economic reforms. Prior to these reforms farmers were required to work in large cooperatives, pooling land, labour and resources to grow state-directed produce that would be sold at set prices. Collectivisation was thought to benefit small urban farmers by lowering individual costs and allowing them to share in the profits derived from economies of scale. Returns were divided among the households according to the number of farming hours members had committed. Since *doi moi* economic liberalisation has allowed farmers to lease their own plots, choose the crops they wish to grow and sell produce themselves at market-determined prices. In addition farmers have greater access to a wider range of seeds and agricultural inputs, such as pesticides and fertilisers, enabling them to farm their plots more intensively.

Most farmers are benefiting from the *doi moi* reforms. Many have managed to secure higher incomes by producing more produce that can be sold at higher market prices, and have diversified away from rice production into more profitable vegetable and fruit production. However these gains have been accompanied by new costs such as land price rises that threaten to cancel urban agricultural leases; uncertainty and variability of market prices; degradation of soils and waterways as a result of increased pesticide and fertiliser usage; and costs, effort and time associated with marketing and trading produce, purchasing equipment and transporting goods, which is now done individually as opposed to collectively.

The shift from collective to individual farming within Hanoi has brought many positives to the city's agricultural community. However many of the benefits of collectivisation have been lost in the process. A more collaborative approach, which allows for collective efforts in some aspects of agricultural production and individualism in others, may provide a more effective form of farming, not just in Hanoi, but for many small farmers in Southeast Asia.

Brody Lee, Department of Geography, University of Otago

Sources and further reading: Jansen *et al.* (1996)

provide low cost, often part-time unskilled female labour forces for industrial activities, and may eventually play an important role in maintaining the environmental qualities of cities through recycling, absorbing organic waste products, and developing alternative energy sources (McGee 1995; Laquian 2005). Economically EMRs are the most powerful centres of the region and raise similar questions to primate cities in terms of uneven development. Some argue that EMRs are the most viable and productive urban forms of the future and should be encouraged to grow in order to drag the region forwards. Others are concerned about environmental degradation, the quality of life of poor people in excessively urbanised areas, and the uneven national development pathways and opportunities that derive from the concentration of wealth, resources and skills into EMRs.

Challenges for urban centres

The class divisions outlined in Chapter 4 are most evident within the compact urban spaces of cities. In the capital cities of the market-led economies the rich and powerful drive Mercedes Benz past the impoverished and homeless, people feast on gourmet foods from around the world while others suffer from malnutrition, and massive mega-malls provide sharp gleaming contrasts to the unserviced squatter settlements nearby. In socialist or post-conflict cities such as Vientiane, Hanoi, Phnom Penh, Yangon and Dili the degree of inequality is less apparent but likely to grow with economic development. Such inequality raises pressing questions in regards to equitable forms of development, three of which will be addressed here: urban migration, employment and housing.

Migration

The ongoing urbanisation of Southeast Asia is driven primarily by the internal migration of people from rural to urban areas (see Figures 6.4–6.5). This migration in turn is driven by uneven development as rural people are attracted by the higher incomes of the city, the improved services in sectors such as health and education, and the modern lifestyles cities offer when compared to rural spaces. Most rural workers find jobs in the informal sector in occupations such as food-stall operators, street vendors, tricycle drivers, garbage pickers or recyclers, or formal sector work as construction workers, security guards, domestic servants or low skilled export oriented factory workers (Rigg 1998). In rapidly urbanising areas, such as those on the island of Batam in the IMS-GT, migrants – many of whom are from

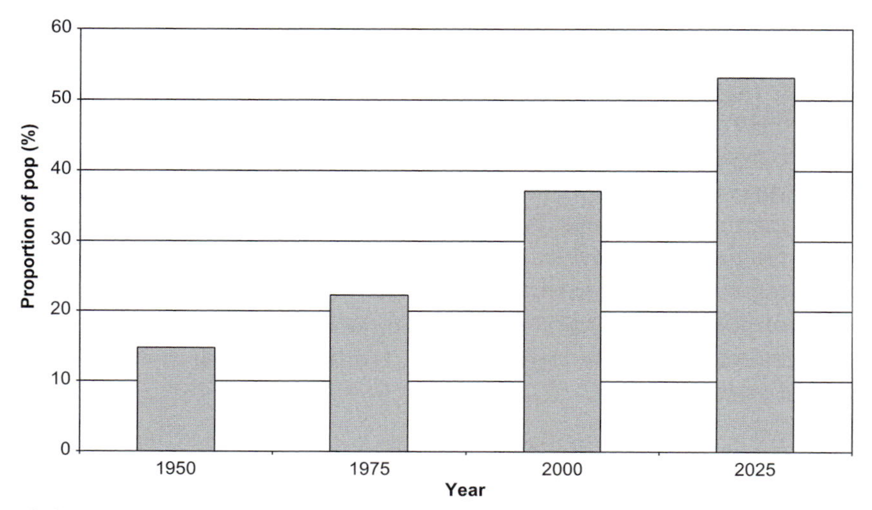

Figure 6.4 Proportion of population living in urban areas in Southeast Asia 1950–2025 (projected).

Source: Jones (2002)

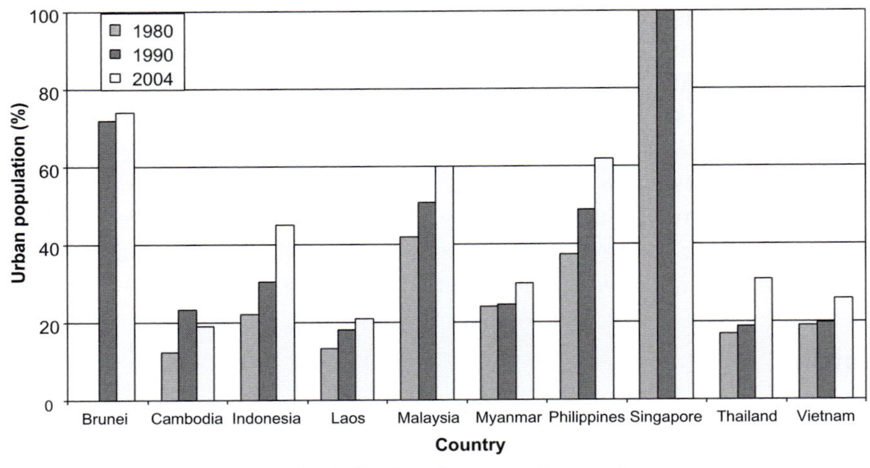

Figure 6.5 Proportion of population living in urban areas by country.

Source: ASEAN Secretariat (2005a)

rural areas – vastly outnumber the original population. However the simple division between rural spaces and urban spaces is becoming harder to maintain given the evolution of *desakota* areas in EMRs as well as ongoing links between urban migrants and rural households. Some migrants move permanently to urban areas but are still very important income earners for their original rural families who they support by regularly sending of home a portion of their income as remittances. Other migrants work for temporary periods, moving from rural areas when they are young adults, for example, and staying for several years before returning home to marry and raise families. There are also seasonal migrants who may work rural fields during harvesting season but migrate to the city during other seasons when rural labour opportunities and demands are lessened. And finally there are those living in rural spaces at the edges of urban settlements or in the *desakota* of EMRs who commute to urban workplaces on improved transport systems on a daily basis. The increasing mobility of people, resources and finances between rural and urban spaces is blurring the boundaries of these traditional spatial divides (see Lynch 2005).

International migration also plays an important role within Southeast Asian cities by filling labour market gaps in receiving countries and providing an additional form of income for the households of source countries. The most sought-after migrants are small numbers of highly skilled and educated migrants that fill managerial or technical roles within the private and public sector. Traditionally such migrants have come from the developed countries such as the US, Britain, Australia and Japan, but they are increasingly coming from South Korea, China and India (Yeoh 2006). They work for temporary periods and earn high wages with considerable benefits in the wealthier labour markets of Singapore and Malaysia, but also within senior positions of TNCs throughout the region. At the opposite end of the market are unskilled or low skilled labourers who come from poorer countries such as Indonesia and the Philippines, and more recently Myanmar, Cambodia and Vietnam, to fill gaps at the bottom end of the labour market. For receiving countries such as Malaysia, Singapore and to a lesser extent, Thailand, low end migrants provide a cheap and flexible labour force that can keep costs down and boost the international competitiveness of urban industries. Perspectives on international migrants differ from those who see them as empowered actors seeking out higher incomes to support themselves, their homes and their homelands to those who focus on the negative aspects. In the Philippines the important role that the international remittances play within household and national economies has been recognised and overseas workers are celebrated as 'national heroes'. Others concentrate upon the difficult working conditions such migrants are faced with which often include poor wages (when compared to other workers in the receiving country), monotonous, isolated, unhealthy or hazardous workplace environments, poor union representation, and visas that restrict employment options and ensure they are employed on a 'use and discard basis'. Domestic workers in Singapore, for example, are not allowed to become pregnant or change employers, must undergo regular medical screenings, and are not included in national labour laws that guarantee at least one day off per week (Yeoh 2006). Such discriminatory policies have led some commentators to argue that low skilled international labour migration is an extreme form of exploitation that naturalises the separation of workers from their homes and families.

Concerns about exploitation have magnified since the 1990s when international labour flows became increasingly feminised. Women now make up the majority of international migrants into urban spaces and are filling labour openings in domestic work, service industries such as hospitality and entertainment, as well as assembly line work in light manufacturing industries. Their attraction to employers is based upon gendered stereotypes about women's reproductive roles, particularly within domestic work as maids or nannies, as well as the gendered constructions of female workers as more docile, subordinate and controllable than men (Devasahayam *et al.* 2004). Reports of verbal, physical and sexual abuse of migrant women, particularly when working in domestic spaces as maids and nannies, is common in countries such as Singapore and Malaysia and a range of NGOs have formed to try and protect their rights. Piper (2004) has observed that such labour activism can work against their employment prospects as less organised Indonesian and Vietnamese domestic workers are becoming increasingly preferred to more organised Filipino workers. As to whether international migrants should be seen as 'heroes' or 'victims', much depends on the individual case, with in-depth studies showing that women display considerable agency in improving their working conditions at the personal scale (see Gibson *et al.* 2001).

Employment

One of the key outcomes of economic development within Southeast Asia has been the expansion of employment opportunities in urban manufacturing and service industries. Many jobs are in the formal sector, which refers to jobs that are recognised and regulated by government. However jobs have also expanded in the informal sector which refers to small scale and/or family-run entrepreneurial businesses that are unregulated by government and require relatively few skills or start up capital. The regulations that protect basic working conditions and rates of pay in the formal sector are absent in the informal sector making it a riskier place to work, however, as will be discussed, there are also advantages to informal sector employment.

Formal sector employment

Within most economic development theories, such as modernisation, neo-Marxist theories and neo-liberalism, one of the key objectives is to provide people with well paying formal sector jobs. If these jobs are in urban areas based on secondary and tertiary industries it is taken as a sign of a maturing economy that is improving people's quality of life and dragging them out of poverty. The rationale behind these approaches is that secondary and tertiary industries provide higher wages for their workers, which equates to greater opportunities and more secure lifestyles. Certainly these claims would appear to be true for those urban workers in elite or middle-class circles who have managed to secure jobs in finance, management, information technology or a range of other skilled positions. The middle classes of

Singapore, Kuala Lumpur, Manila, Bangkok and Jakarta have access to a wide range of consumption items, services and recreational experiences that mimic or surpass those of developed cities around the world. However, with the exception of Singapore, urban middle-class jobs are still relatively limited and only accessible by those who are well educated and/or well connected. Even in Singapore university graduates face considerable hurdles in acquiring prime positions because of the competition they face from skilled migrants emigrating from international labour markets (see Appold 2005).

Many formal sector workers in Southeast Asia find jobs in working-class positions such as work in labour-intensive textile, electronics or footwear factories, chemical or heavy machinery industries, construction, and basic clerical or sales positions. Whether development is benefiting the quality of life of some of these workers is the subject of anti-globalisation and anti-sweatshop debates that are critical of the low pay and limited opportunities for the urban poor. Anti-globalisation critics focus on the poor working conditions these formal sector workers are exposed to which can include long hours, little or no reward for overtime, monotonous, repetitive and menial daily work activities, dangerous or hazardous working conditions, verbal and sometimes physical harassment and exploitation, and little financial return for their labour. Anti-sweatshop debates focus on the horror stories that abound about the poor treatment of factory workers where people undertake repetitive work for 6–7 days per week in difficult conditions while earning little more than a dollar a day (see Klein 2000).

These poor working conditions have underpinned the successful EOI strategies that have propelled national scale economic development in the region. To attract FDI the cost of producing goods must be kept low, which requires a cheap labour force. Hence working class wages and labour conditions within many urban centres of the original ASEAN-5 nations have not improved at the same rate as the national economy, and in some places barely seem to have improved at all. This reflects negative government and employer attitudes towards labour unions (see Chapter 5) as well as ongoing migration to urban centres that fills labour gaps in growing economies, suppressing pressure for wage rises. With increasing competition for FDI from China and greater intra-regional competition from Vietnam, Laos and Cambodia, which offer some of the lowest wage rates in the world, the 'race-to-the-bottom' scenario feared by many anti-globalisation advocates may well be becoming a reality for some with working conditions for the urban poor not expected to rise dramatically any time soon (see also Jenkins 2004; Nadvi *et al.* 2004; Mosley and Uno 2007).

In contrast neo-liberal economists are more likely to stress the positive aspects of working-class occupations. Despite the difficulties of factory work, for example, most people who move to cities still see this as better than the alternatives, which is often poorly paid and insecure informal sector jobs or a return to their rural village. Indeed employment in a factory can be seen as a sign of prestige and advancement for many rural migrants. It allows some women to break free of restrictive gender roles within their families and experience new worlds and opportunities, and it can also be seen as a rite of passage for young adults before they return home and raise families. For temporary migrants the poor working conditions are worth enduring for the financial returns, which allow them to support their rural households while away while also returning with savings that can be invested in land, vehicles, consumer goods or productive investment activities. Other workers use their incomes to seek further education that will assist them in seeking better quality jobs. While this positive view of urban factory work is backed up by the constant demand for urban employment in factories it should not be used to condone the exploitative conditions within such workplaces. In the absence of national scale industrial reform in-depth studies have revealed the dissatisfaction workers feel about their conditions and the strategies they are adopting to improve their workplaces. In a seminal study Ong (1987) documented how female factory workers in Malaysia frequently experience evil spirits in their workplaces which cause them to damage their equipment and attack employers, forcing temporary closure of factories so they can be purified, all acts that can be read as responses and resistance to difficult workplace situations. Elias (2005) notes other forms of 'everyday resistance' within Malaysian factories that include high rates of absenteeism and a lack of loyalty to employers that allows them to shift to factories where conditions may be improving. While these localised forms of resistance and protest may lead to temporary respites from workplace pressures substantial improvements to labour relations would appear to lie in building union solidarity and extending links to international networks and institutions such as the International Labour Organisation. Until working conditions improve anti-sweatshop debates and movements will continue to try and improve the quality of life of those employed in urban factories (see Box 6.2).

Box 6.2

Urban manufacturing industries: the sweatshop debate

One of the main drivers of Southeast Asian economic development has been the region's ability to attract FDI into low skill, labour-intensive urban and peri-urban manufacturing industries. However over the last fifteen years these industries, typically in textile, garment, footwear and electronics sectors, have come under intense criticism from labour rights NGOs and from the mainstream media:

> Yati sits at a sewing machine, which is one of sixty in her row. There are forty-six rows on the factory floor. For working sixty-three hours a week, Yati earns not quite $80 a month – just about the price of a pair of Reeboks in the United States. Her hourly pay is less than 32 cents per hour, which exceeds the minimum wage for her region in Indonesia. Yati lives in a nearby ten-by-twelve foot shack with no furniture. She and her two roommates sleep on the mud and tile floor.
>
> (*Boston Globe* 10 July 1994, cited in Miller 2003: 102)

Yati's story is not uncommon and labour rights NGOs accuse TNCs of regularly abusing workers' rights in pursuit of foreign profits. Common allegations include: bans on union representation; workers not being able to earn a liveable wage; controlled use of bathroom breaks; verbal and sexual abuse; punishments for non-work related incidents; inadequate health care and dangerous exposure to workplace hazards; forced overtime to meet extremely high work quotas; and not being paid overtime if quotas are not met. Such workplaces are often referred to as sweatshops and have been the subject of bestselling books such as Naomi Klein's (2000) *No Logo*.

Although few would dispute the fact that the pay and working conditions of those producing goods for TNCs in factories across Southeast Asia are far worse than the conditions of those in the West, who consume the bulk of these goods, there are ongoing debates about the benefits of such industries. Research suggests that such jobs are often considered desirable by local workers because the low pay is often better than what unskilled workers would receive in the informal industry; their income is secure; they can be stepping stones to better jobs; they open up new opportunities for those wishing to escape the poverty and tradition of rural areas, particularly women; and they expand worker's consumption habits (see Foo and Lim 1989). For national economies they help absorb excess labour while generating sources of foreign revenue.

A global 'sweatshop debate' is currently underway and it contrasts the pros and cons of such industries. Miller (2003: 103) has argued that exploitation should be defined by workplace relations not by comparisons with locally relevant alternatives,

> If workers are denied the right to organise, suffer unsafe and abusive working conditions, are forced to work overtime, or are paid less than a living wage, then they work in a sweatshop, regardless of how they came to take their jobs or if the alternatives they face are worse yet.

The global anti-sweatshop movement argues that one way consumers can contribute to the minimisation of workplace abuses in Southeast Asian industries is to insist on union labelling and only consume products made in unionised factories.

Sources and further reading: Powell and Skarbek (2006), Bradsher (2006), Klein (2000), Foo and Lim (1989), Miller (2003)

The economic divisions within urban formal sector employment are becoming increasingly spatialised and gendered within urban centres. In EMRs, for example, most of the high paying middle-class occupations are located in the expensive core zones of those cities that are the centres of finance, commerce, retail, service and governance. Older light industries, such as garment, textile or footwear industries, as well as electronic industries, surround core areas in the metropolitan zone while newer heavy and chemical industries have taken advantage of improved transport infrastructure, cheaper land and labour, and city incentives to decentre their activities to the extended metropolitan zone. It is in the outer rings of industrial activity located far from the city centre that newly arrived rural migrants, as well as the unskilled urban poor, find employment. This is leading to a spatial distribution of labour with those working in high skilled middle-class positions having little contact with those in working-class areas. It is also leading to gender segregation with women being preferred in service and light/electronic industries, partly because of their greater dexterity and 'nimble fingers' (see Rigg 2003: 254), leading to women being over-represented in the inner areas of EMRs. In contrast men are more likely to find work in heavy industries in the extended metropolitan zone. Nakawaga (2004) has found the gender ratio in the Bangkok Metropolis of women to men to be 100:85 in the most productive age group (25–49), and while many women may be temporary migrants to the centre it does raise important questions about marriage rates and changing household compositions. These spatial divisions emphasise the unevenness of development within Southeast Asian cities, something that is further evident within informal sector employment.

Informal sector employment

Under conventional economic development theories informal sector employment is expected to decrease as the formal sector expands. The informal sector has traditionally been seen as a remnant of earlier societies with little to offer modern developed economies. Contrary to expectations, however, the informal sector retains an important role within Southeast Asia and employs millions of people in urban areas. The shift towards more human-centred development theories has led to a reappraisal of the informal sector that recognises its core role in sustaining urban livelihoods. Indeed Latouche (1993), writing from a post-development perspective, has argued that it is within informal sectors that alternatives to conventional development are being explored and pursued, and as such the sector should be encouraged to continue. Grassroots theorists, in believing development should be driven by people rather than economic models, also seek to establish ways of improving the lifestyles of those in the informal sector that do not necessarily entail a shift into formal sector employment. Even conventional development economists have come to recognise the importance of the informal sector not only in absorbing excess labour but also in providing important goods, transport and distribution services that sustain the formal economy.

Typical informal sector activities in Southeast Asian cities include small scale transport services, such as cyclo/rickshaw drivers; street vendors, such as food stalls, sidewalk barbers and shoe shiners; as well as porters, handicraft makers, carpenters, builders, beggars and sex workers who work outside the laws and regulations of the state. The informal sectors of Southeast Asian cities provide them with a lot of their uniqueness and personality. It is hard to miss the cyclo drivers of Ho Chi Minh City, the pirate CD hawkers of Bangkok, or the amazing array of food stalls found throughout the region from the baguette stalls on the side-streets of Vientiane, the *tempeh*, *mie goreng* and *murtabak manis* street sellers in Jakarta and *pad thai* or *tom yum goong* food stalls of Bangkok. While informal sector workers can be found in all urban centres of Southeast Asia, they are particularly common within places that have small or relatively undeveloped formal sectors and their numbers swelled when formal centre occupations shrunk during the Asian economic crisis. Informal employment is on the rise in the transitional economies of Vietnam and Laos, where they have relaxed laws that previously constrained such entrepreneurial activities, as well as in the recovering post-conflict states of Cambodia and Timor-Leste. In Cambodia the International Labour Organisation estimates 85 per cent of the country's workforce is involved in the informal economy generating approximately 65 per cent of Cambodia's GNP (Monyrath 2005). Another International Labour Organisation report has estimated that informal women workers in Thailand's urban sex industries are responsible for remitting US$300 million a year back to their rural households, surpassing the budgets of many government sponsored development programmes in the country (Lim 1998).

By definition informal economies have grown outside of government policies and legislation, raising the question as to whether governments, or any other development actors, should become involved in the future. The main arguments for external intervention or regulation revolve around issues of safety and labour rights. With a lack of government regulations or unions there is the potential for people to be exploited or exposed to hazardous situations without recourse to any official complaints or protection system. An unregistered cyclo driver in Vietnam who has an accident, for example, is unlikely to receive any significant compensation from the cyclo owner. Similarly there may be no official guidelines relating to food preparation, which may, on occasion, make food unsafe for eating or expose cooks to unsafe kitchen environments. There are also few official mechanisms workers can turn to if they are abused, harassed, underpaid or otherwise exploited by their employers. While some Southeast Asian governments do not necessarily have a good record of protection these rights in the formal sector there is at least greater potential for official action there than in the informal sector. The main form of state action has come through criminalising certain elements of the informal economy such as prostitution, drug trafficking and, in some places, begging. Outside of criminalisation, however, informal businesses are mainly monitored through community and kinship networks rather than state surveillance. Many informal economies are based on social capital, the informal relationships and networks that exist at the community scale that are vital to small business success. If word gets out that someone engages in exploitative or unscrupulous activities his or her social capital can be weakened or lost, encouraging self-policing and fair business principles. To further improve the informal sector many international NGOs, inspired by grassroots theories, have joined domestic civil society organisations in funding programmes that strengthen the resilience of this sector by providing access to micro-finance, skills training and equipment.

Housing

A basic measure of successful equitable development at the national scale is the provision of safe and adequate housing to a country's population. In Southeast Asian cities housing quality varies dramatically between and within urban spaces, and many places have failed to provide adequate housing for all. Singapore, with its unique landscape of high rise residential apartment blocks, is the exception having found a high altitude solution to its space shortages (see Figure 6.6). Indeed the greatest concerns for urban Singaporeans in terms of housing, apart from affordability, relate to issues of lift breakdown and lift crime as buildings escalate ever upwards (Yuen *et al.* 2006). Elsewhere, in the sprawling EMRs of the region there is a far greater range of housing types, from elites occupying luxurious bungalows in gated communities, to middle-and working-class families occupying a range of low and high rise dwellings of varying value and convenience, to sprawling squatter settlements that have arisen to house the urban poor. Diversity is most pronounced in market-led economies but is growing in socialist cities such as Hanoi, Ho Chi Minh City and Vientiane due economic liberalisation and a relaxation

Figure 6.6 High rise living in Singapore.

Source: Author

on laws governing domestic migration (Forbes and Thift 1984; Ha and Wong 1999). Cities in post-conflict states such as Timor-Leste and Cambodia have fewer wealthy areas but large squatter settlements as a consequence of the violence that left many homes destroyed and unliveable. The United Nations estimates, for example, that approximately 70 per cent of the public infrastructure was destroyed and the majority of urban houses were damaged or left in ruins when Indonesian militias went on a rampage following Timor-Leste's vote for independence in 1999 (see Figures 6.7–6.9).

One of the reasons why there is such housing diversity within the wealthier cities of the region relates to land tenure issues. The rapid historic growth of the ASEAN-5 economies brought new levels of wealth to urban elites and the emerging middle classes. Rather than invest this new money in local businesses, where there were limited opportunities due to the domination of foreign capital, wealth was invested in land and property to raise incomes and boost local reputations. Land speculation suited the newly rich as it required a minimum of business savvy, had limited opportunities for corruption and was locally based and tangible. The demand for urban property

Figure 6.7 Destroyed buildings in Timor-Leste.

Source: Author

Figure 6.8 Destroyed building in Dili, Timor-Leste.

Source: Author

quickly escalated as economies boomed eventually pushing land and housing prices above those of many developed countries (see Evers and Korff 2000). These rising land prices put home ownership out of the price range of rural migrants moving into cities as well as restricting access for working classes and lower middle income earners. This has led to frustration and in some cases desperation for the urban poor as well as huge profits for wealthy landowners. While these problems were initially avoided in socialist states where private land ownership is illegal, the recent relaxation of land laws that allows people to lease and trade leases is resulting in soaring prices in the booming economies of Hanoi and Ho Chi Minh City.

The expensive price of land has contributed to extremes of housing quality between rich and poor. Wealthy residents are increasingly choosing to locate themselves in gated communities that are fenced off from the broader population with guards limiting access to those approved by residents. These settlements began forming as early as the 1960s in Manila, the 1970s in Jakarta and in other Indonesian centres such as Surabaya by the 1980s (Dick and Rimmer 1998). There are two types of gated communities, those that are planned by city authorities

Figure 6.9 Hopeful graffiti in Dili, Timor-Leste.

Source: Author

and urban property developers, and those that occur spontaneously. Jakarta's new towns are examples of planned gated communities that have been built both on the fringes of the city (but close to transport infrastructure) and within more central urban spaces to house high and upper middle income groups in secure and luxurious lifestyles. The new towns boast a range of high class services and facilities such as self-contained golf courses, equestrian areas, shopping malls and hospitals that are only accessible to residents and their guests (see Firman 2004a). Spontaneous gated settlements are unplanned and instead arise from the collective action of wealthy communities concerned about their safety. These are particularly common for Chinese communities whose wealth has been the target of anti-Chinese frustration and aggression in the past. However even these structures do not necessarily prevent violence as the Jakarta riots of May 1998 showed when over 1,000 houses and 5,000 commercial buildings were damaged, many of which were situated within gated communities (Hun 2002). Despite failing to protect residents the riots have increased demand for more secure gated communities as wealthy residents seek to keep themselves shielded, socially, physically and economically, from the difficulties experienced by those who surround them.

At the other end of the scale are rural migrants who have moved to the outer zones of urban spaces in search of factory employment as well as urban refugees fleeing the rising prices of the core (see Murakami *et al.* 2005; Firman 2004b; Jones 2002). Despite government initiatives to provide public housing for low income people the sheer volume of urban migrants means it is unlikely governments will ever catch up with demand. With few alternatives available most end up in informal housing settlements in which they use low cost local materials to illegally build shelters on public or privately owned land. In most cases they have no security of tenure placing them in precarious legal situations should the state or private landowners decide to evict them. Their illegal status means residents are likely to have difficulty accessing basic urban services, such as electricity and water provision or basic sanitation and rubbish removal. These informal settlements, referred to as slums, squatter settlements or shanty towns, have become semi-permanent features of Southeast Asian urban landscapes, providing a crucial role in housing poorer populations. The United Nations Human Settlement Programme (2003) estimates 28 per cent of all urban dwellers in Southeast Asia live in these slum settlements. Attitudes towards such spaces vary from those who portray them as places of crime, disease and poverty, to more recent perspectives that recognise the crucial role they play in urban economies. Subsequent policies also differ from strategies based on eviction and/or relocation, to strategies based on development and improvement.

The important role of informal settlements can be seen in Metro Manila where they house 25–40 per cent of the city's residents (Porio 2002; Shatkin 2004). Many of the inhabitants have come from rural areas but many are also refugees from urban development. Between 1997–2000, for example, Urban Poor Associates estimate more than 300,000 people were evicted from their homes within Manila to make way for new rail lines, office spaces and roads (Shatkin 2004). Most of the informal settlements are in the extended metropolitan zone although some are located on hazardous inner-city land that is unwanted and unused by other developers. Those on hazardous land face threats of mudslides and flooding, which are frequent and dangerous events in Filipino cities. Squatters face a number of other challenges such as maintaining basic sanitation and hygiene with few waste management services and accessing things such as water. Aiga and Umenai (2002) report that women and children, as part of their gendered reproductive duties can queue for hours a day to collect water from public faucets. Probably the single biggest issue, however, is securing land tenure so that they can feel confident about investing in their properties. Without security of tenure there is very little impetus to build secure structures that could, at any time, be legally demolished to make way for urban development. A number of civil society organisations have formed to lobby authorities to grant formal settlement rights to squatters with varying degrees of success (see Porio 2002; Porio and Crisol 2004). One positive response from the government has been to encourage settlers to buy public land from the state by offering mortgage loans that the settlers pay off over time (see Box 6.3). These types of innovative initiatives are required to rectify the poor housing conditions many squatters live in and suggest ways of working towards more inclusive, empowering and equitable futures.

Box 6.3

Squatter settlements in Manila

> Children scavengers like us go to the dump site every day after school, and all day on Saturdays and Sundays. We bring a rake, a sack and boots. Also food. We usually wear dirty pants, long-sleeved shirts and hats. On weekdays, we can collect enough to earn P 20–30 and, on weekends, P 80 each day. If we don't go to the site, we won't have an allowance for the next week at school. Some of our parents also work at the site. We like scavenging because we earn from it. Sometimes, accidents happen, like when you get cut by the rake. When that happens, I bandage the wound with a clean strip of cloth. The house of a classmate of mine, also a scavenger, was buried during the landslide and he died. His whole family died, too. I was very sad.
>
> (Child squatter in Quezon City, the Philippines, cited in Racelis and Aguirre 2002: 105)

The above quote outlines some of the hazards facing the urban poor in the Philippines. Unable to secure formal housing they have turned to large scale illegal 'squatting' on unserviced land that is often prone to hazards such as disease, flooding and landslides. Most work in the informal sector, with some of the poorest and least skilled being forced to scavenge dumpsites to make a living. While squatters are politically and economically marginalised by poverty they make up a very significant proportion of the population. There are 7 million squatters in the Philippines, half of whom live in Manila, making up a third of the city's population.

One of the greatest concerns for squatter communities is the constant threat of eviction due to a lack of tenure on the private or public land in which they are squatting. This prevents them from investing in and improving their properties as they live in fear of demolition. These fears have led to the formation of local Resident Associations which have become central to community identity and everyday life. Resident Associations empower squatters by providing services in the absence of the state, representing their interests collectively when bargaining with authorities, mobilising communities to undertake protest actions, forming links with other associations and civil society allies, and blocking moves to evict them. If eviction is inevitable they can still assist communities by helping them access financial compensation and/or participate in the selection of appropriate relocation sites.

Combined action by Residents Associations, civil society NGOs and the government has seen the development of the Community Mortgage Programme (CMP). Under the CMP the government provides squatters with access to formal credit with which they can buy the land on which they reside. The benefits of such schemes are twofold – the state and private land owners can gain access to the capital locked up by illegal settlements, while squatters can gain security through securing land tenure. However, this process still excludes those, like the scavengers described above, who may not be able to meet the financial requirements of servicing a debt. Richer squatters tend to benefit most from such schemes as they are able to buy the mortgages of the very poor to become small scale landlords while the poor remain landless. This has led to conflict within communities between those who gain from the CMP and those who continue to be excluded because of poverty.

Sources: Berner (1997, 2000), Racelis and Aguirre (2002), Porio and Crisol (2004)

Conclusions

Urban centres of Southeast Asia have attracted disproportionate amounts of wealth, resources, services and opportunities from the ongoing economic development of the region. Many of these benefits have accrued to a relatively small number of cities that are generally capitals of their countries and have achieved global, regional or national economic and political significance. However even within these shining symbols of development inequality is rife and it is possible to think of cities within cities. For those with good education and good connections Southeast Asian cities represent worlds of opportunity with well paid jobs located in the central business districts, luxurious housing possibilities and access to world class services and recreational activities. For others, however, many of whom are recent migrants, cities can be spaces of hardship with formal jobs difficult to come by, poorly paid and monotonous, and housing conditions insecure, dirty and/or dangerous. These two groups occupy different spaces in the city and rarely come in contact with one another exacerbating class differences and emphasising the challenges facing more equitable development patterns. At the national scale governments must decide whether they want to slow the rapid rates of urbanisation occurring across the region by redistributing wealth, resources and employment opportunities more evenly across the country or whether they should pursue national economies dominated by megacities and EMRs, which will be responsible for attracting wealth to the nation. At the city scale urban authorities are faced with challenge of improving the working and living conditions of the urban poor and ensuring they have access to a greater proportion of the benefits of development. Solutions lie in a combination of insightful government legislation, organised labour union representation, and collective action, and in national and international civil society networks adopting innovative approaches that empower low income communities. In enhancing the skills and opportunities available to poorer communities in urban spaces more developed, diverse and equitable cities can evolve.

Summary

- **Market forces are concentrating wealth and power into a small number of Southeast Asian cities, some of which have become influential nodes within global economies.**
- **Cities provide a wide range of income earning opportunities that are attracting large numbers of rural migrants leading to the evolution of mega-cities and extended mega-urban regions.**
- **Inequality is clearly evident in income, housing type and employment opportunities within urban spaces**
- **The urban poor are finding employment in informal economies and living illegally in long-established, but under-serviced, squatter communities.**
- **It remains to be seen if socialist states can overcome urban inequalities as they embrace broader market forces.**

Discussion questions

1. Discuss the pros and cons of pursuing national scale development policies based on concentrating investment in one or more urban centres.
2. Compare the strengths and weaknesses of seeking out work in formal and informal labour markets.

3 Discuss the strategies you would adopt to help Southeast Asian cities develop in more equitable ways.

Further reading

For an accessible introduction to urban spaces in Southeast Asia see:

Forbes, D. (1996) *Asian Metropolis: Urbanisation and the Southeast Asian City*. Melbourne: Oxford University Press.

More theoretical explorations include:

McGee, T. and Robinson, I. (eds) (1995) *The Mega-urban Regions of Southeast Asia*. Vancouver: UBC Press.

Evers, H. and Korff, R. (2000) *Southeast Asian Urbanism: The Meaning and Power of Social Space*. New York: St Martin's Press.

An interesting and engaging in-depth study of a Southeast Asian city is:

Askew, M. (2002) *Bangkok: Place, Practice and Representation*. London: Routledge.

For sweatshop debates see:

Foo, G. and Lim, L. (1989) Poverty, ideology and women export factory workers in Southeast Asia. In Afshar, H. and Agarwal, B. (eds) *Women Poverty and Ideology in Asia*. London: Macmillan, pp. 212–233.

Miller, J. (2003) Why economists are wrong about sweatshops and the antisweatshop movement. *Challenge* 46(1): 93–122.

Klein, N. (2000) *No Logo: Taking Aim at the Brand Bullies*. New York: Picador.

Useful websites

The organisation that concentrates on access to shelter in urban centres is UN-Habitat at: *www.unhabitat.org/*

Sweatshop campaigns can be found at: *www.sweatshopwatch.org/*

7 Transforming rural spaces

Introduction

When people outside the region think of Southeast Asia they often think of it as a place steeped in rural tradition. Paddy farmers are imagined to be labouring with water buffalos under hot suns to grow crops that will be harvested by communities in ways that have changed little over generations. While this type of Asian rural idyll is still very relevant to parts of modern Southeast Asia, it is underlain by processes of transformation that are just as significant, albeit often less visible, than urban changes. For example, farmers are much more likely to be growing crops for sale rather than for household consumption and the varieties grown are likely to have been bred by global agribusiness corporations rather than derived from local sources. Fields are inundated with chemical inputs from pesticides, fertilisers and herbicides, and are likely to be owned by distant wealthy landowners rather than those working the fields. People are pursuing new non-farm income earning opportunities; seeking work in the factories that have sprung up on the improved roads that lead to town, or embarking on their own small scale enterprises within rural communities and using cash incomes to buy food and other goods. Still others are leaving rural spaces altogether, attracted by the bright lights, higher incomes, improved services, and modern lifestyles afforded by local and foreign cities.

The rural idyll tends to distract attention from the widespread poverty that encompasses the lives of many living in rural spaces. The agrarian sector has not performed well when compared to the economic vitality of urban manufacturing and service sectors and as a result those living in rural spaces, which comprises the majority of Southeast Asians, have not benefited to the same degree as urban folk from the improved economies and technological achievements of regional development. As a consequence rural spaces have long been the focus of development interventions that try to eradicate poverty, expand livelihood opportunities and improve access to transport and services such as health and education. This chapter will focus on some of the strategies and transformations taking place as a consequence of development processes. The first section will analyse transformations in the agrarian sector, and the second will look at the phenomenon of de-agrarianisation.

Agrarian transformations

The conventional approach to developing rural spaces has concentrated upon transforming and improving the agricultural sector. The importance of this sector to the national economy of Southeast Asian countries varies widely (see Figure 7.1). As a country develops and diversifies its economy it becomes less reliant on primary

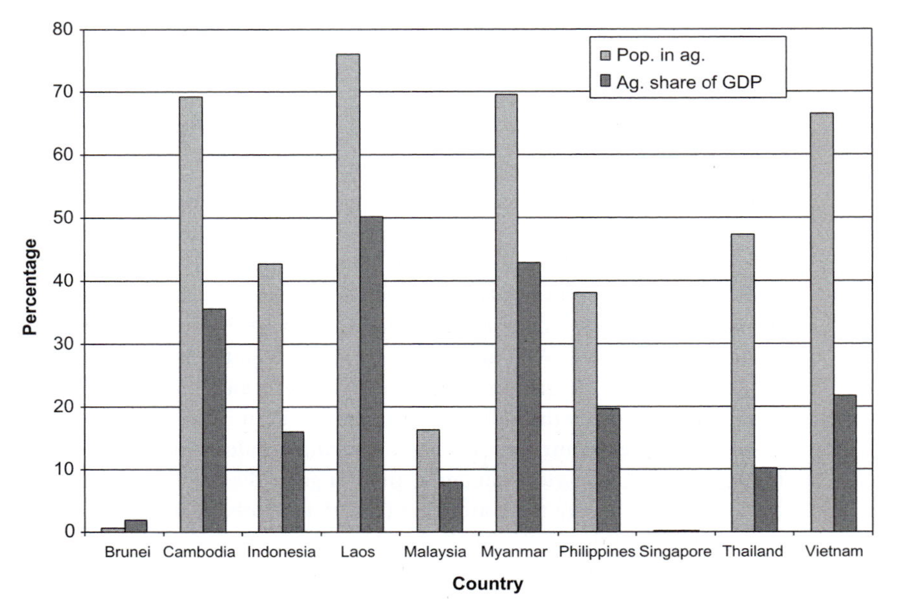

Figure 7.1 ***Proportion of ASEAN population involved in agriculture vs GDP income from agriculture.***

Source: ASEAN Secretariat (2005a)

industries such as agriculture. Hence Vietnam, Laos, Cambodia and Myanmar are more reliant on agriculture than the more developed ASEAN-5 economies that have advanced manufacturing and service sectors. This reflects the earlier centralised command economies of socialist states that saw them invest comparatively high proportions of national income into the development of rural spaces. These structures are particularly strong in Vietnam and have been an asset in its recent battle to combat the avian flu virus. In contrast the privatisation of government services and ongoing decentralisation policies are detracting from Indonesia's ability to combat this threat (see Box 7.1). Most rural people are employed or otherwise reliant upon agricultural development and positive agrarian transformations should bring substantial improvements to rural quality of life. The most commonly grown crop produced in the region is rice, which is either produced for household consumption, known as subsistence agriculture, or produced for sale and exchange to local, national or international markets, known as commercial agriculture. Traditional development theories have encouraged a shift from subsistence to commercial agriculture for a variety of reasons. These include the need to bring cash economies to rural areas to enable farmers to purchase goods and services from other places; to facilitate trade and exchange with

Box 7.1

Combating avian flu virus H5N1 in rural spaces of Southeast Asia

The avian flu virus H5N1 is the latest public health scare to affect Southeast Asian region. In 2003 SARS spread through the region with Singapore (238 people infected, 33 deaths) and Vietnam (63 infections, 5 deaths) the worst affected. The avian flu virus H5N1, which first crossed the inter-species boundary from birds to humans in Hong Kong in 1997, had its first major outbreak in Southeast Asia in 2004. It spread through the Vietnam and Thai poultry industries before rapidly infecting Indonesia and, more recently, Cambodia. Currently the only humans that have been infected are those who have had some exposure to sick birds, often rural poultry workers. However, it is apparent that the virus is mutating and if it gains the ability to be passed from human to human there are grave fears that a global pandemic will take place. Consequently the World Health Organisation has been focusing on combating the problem in Southeast Asia where the bird–human transmission rates have been high. By August 2007 Indonesia had 105 cases with 84 deaths, Vietnam 95 cases/42 deaths, Thailand 25/17, Cambodia 7/7, and Laos 2/2. The virus has since spread on migrating birds as far as Africa and Europe.

The approaches adopted in different Southeast Asian countries reflect their governance styles and the quality of their rural health and veterinary services. In Northern Vietnam H5N1 first jumped to humans in January 2004 where it infected 87 people and killed 38 within 18 months. The central government opted for mass top-down poultry vaccination and culling programmes, claiming they would eventually protect all birds from the virus and stop its spread to humans. The government also stated in late 2005 that it had reduced the death rate in humans from around 70 per cent to 20 per cent through the use of anti-viral drugs provided through its national rural health system. However, it soon became clear that Vietnam did not have the rural health resources to undertake such a large scale programme without outside help and expertise. More worryingly a World Health Organisation inspection of the hotspots in Northern Vietnam found that new strains and symptoms had emerged leading to mis-diagnoses and sick people being sent back into the community while still being potentially infectious. However the mass vaccination and culling programme has limited new human cases to just two since 2006, and is being hailed by most observers as a successful means of slowing the spread of the virus.

In contrast rural health and veterinary services have been cutback and localised in Indonesia as a consequence of neo-liberal policies and decentralisation. In Maluku province, for example, there were once five government funded vets but they have all left for higher pay in Java since decentralisation. This has hampered any attempts at introducing national top-down vaccination and culling programmes and while local authorities are doing their best, a lack of resources and coordination has contributed to the spread of the virus to thirty of Indonesia's thirty-three provinces. With limited national capacity and oversight, underfunded public health programmes, and 60 per cent of all Indonesian households having chickens in their backyard, some 300 million birds, the conditions are ripe for the virus to continue to spread and mutate. The Indonesian government is working with Food and Agriculture Organisation (FAO) and World Health Organisation to try and contain outbreaks but with limited national and international funding, there is a long way to go.

Sources: World Health Organisation (2007a, 2007b), Normille (2007), Kristiansen and Santoso (2006)

urban dwellers that allows them to pursue non-agriculture pursuits rather than have to grow food themselves; and to boost national GDP and raise foreign exchange earnings through sales on international markets. Most Southeast Asian farmers, particularly those located close to transport routes, are involved in commercial agriculture, which has lead to a diversity of cash crops being grown that include sugar, corn, cassava, tobacco, coffee, rubber and, more recently, oil palm (see Figure 7.2). While smaller grassroots development initiatives

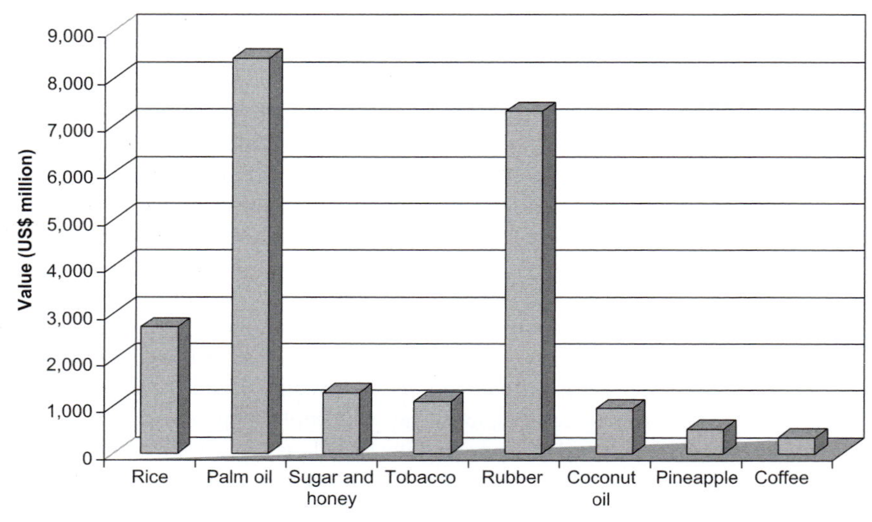

Figure 7.2 ***Value of major ASEAN crop exports 2004.***

Source: ASEAN Secretariat (2005a)

concentrate upon sustaining the livelihoods of those that practice subsistence agriculture the majority of effort has been directed towards increasing the productivity and profitability of commercial enterprises. The three types of agricultural development initiatives that will be reviewed here are agricultural expansion, agricultural intensification and rural land reform.

Agricultural expansion

The simplest way of increasing agricultural economies is to expand the volume of land dedicated to agricultural production. Small scale agricultural expansion happens spontaneously when farmers decide to cut into forested land or utilise swamp or dry lands to expand their production. This unplanned form of agricultural expansion is often driven by poverty or population pressures and can result in fragile, marginal or protected areas being damaged or exhausted by farming. While spontaneous agricultural expansion is often frowned upon by authorities, and may be illegal if it takes place in protected zones, planned agricultural expansion is a core development policy within rural spaces. One of the largest, longest running and most successful programmes in the region is that which has been coordinated by the Malaysian Federal Land Development Authority (FELDA) since 1956. The programme assists landless peasants, who apply for selection, to move to recently cleared locations where they develop new communities and economies based on cash crop plantations. The scheme provides technical support for new migrants as well as land, houses, schools and clinics in new *kampongs*, with only a portion of the start up costs having to be paid back once the plantations become profitable. Migrants pool labour, land and resources in the early years of the programme while the state provides good transport infrastructure and local processing facilities to enhance the competitiveness of the new farms. In recent years the prime crop has changed from rubber to oil palm for which Malaysia is the world's largest producer. The agricultural enterprises that result are often referred to as forms of industrial agriculture because of their large scale and reliance upon industrial inputs such as pesticides, herbicides, fertilisers and cropping machinery such as tractors. While having clear benefits for those involved in the programme, as well as the national economy, the programme has attracted criticism for its cost, the lack of non-Malay ethnic groups participating in the scheme, the widespread clearance of forests that has accompanied the programme, and a recent shift away from granting land to granting shares – thereby perpetuating landlessness. This shift in policy has resulted from a controversial market-oriented part-privatisation of the programme where relocated farmers no longer receive land titles, but become shareholders of large private estates on which they work. Despite its faults FELDA represents a large scale rural development programme that has succeeded in improving the lives of many of Malaysia's rural poor (see Ghee and Dorall 1992).

The single most ambitious agricultural expansion programme attempted in Southeast Asia, and perhaps in the world, has been Indonesia's World Bank funded *transmigrasi* programme. The programme was initiated in the colonial period but continued after independence with a primary aim of relieving population pressures in high density areas. Most migrants are sourced from Java, where approximately 60 per cent of Indonesians reside, but also Bali and Lombok, and helped relocate to less populated parts of Indonesia's sprawling archipelago. *Transmigrasi* hoped to bring modern agricultural techniques and rural development to the less developed Outer Islands but it was accompanied by an additional nation-building motive. By sending migrants to remote areas the programme sought to build a national identity through 'Javanising' marginal areas and developing a more unified national identity (see Elmhirst 1999). One of the early Ministers for Transmigration is quoted as saying:

> By way of Transmigration, we will try to realise what has been pledged: To integrate all ethnic groups into one nation, the Indonesian nation. The different ethnic groups will in the long run disappear because of integration and there will be one kind of man, Indonesian.
>
> (Cited in Hoey 2003: 112)

Given these multiple objectives considerable resources have been invested into the programme and well over a million families, some 5 million people, lured by promises of transport, land, financial incentives and agricultural support, have been relocated. Some of the more common destination areas included Irian Jaya, Sulawesi, Sumatra and Kalimantan.

There were three main transmigrant types: sponsored transmigrants, who made up the bulk of the programmes and received transportation, land, a small house and other social services; spontaneous transmigrants who received the same package but were required to pay for their own transportation; and local transmigrants who lived near transmigrant settlements but could receive similar packages (Leinbach and Smith 1994). Less publicised programmes included those for ex-military personnel who were encouraged and assisted to transmigrate to troubled areas such as East Timor and Irian Jaya (Fearnside 1997). Unlike the FELDA programme, which has always been oriented towards industrial agriculture, early transmigrants were given small 1–2 hectare farms on which they were encouraged to grow paddy and other crops that could sustain household incomes above subsistence rates. Later transmigrants, however, have been required to grow oil palm on large plantations or harvest wood from industrial forest plantations and have far less autonomy and control over the land. Not surprisingly, given the scale of the programme, *transmigrasi* has attracted a range of criticisms that include:

- A lack of real impact of Java's population density issues.
- A lack of education and understanding of local ecosystems and agricultural processes, particularly among urban transmigrants, leading to poor yields, hunger and hardship in transmigrant communities.
- Poor yields and financial returns necessitating a diversification of household income sources, particularly to off-farm employment and income (see Leinbach and Smith 1994).
- Conflict with existing indigenous communities over land rights, access, resources and cultural differences and local resistance to

modernised agricultural techniques (for local resistance see Elmhirst 1999).
- Widespread ecological damage of forest clearance leading to Indonesia's high rate of deforestation (see Chapter 8).

These criticisms, combined with the demise of Suharto's New Order government, saw the programme officially completed in 2000, however many believe it continues in modified forms today.

Agricultural expansion has also taken place in the socialist states of Southeast Asia. Vietnam, for example, encouraged over a million people to leave Hanoi and its immediate surrounds to settle in sparsely populated new economic zones north of the city. This movement was driven by Malthusian concerns regarding the ability of the rural areas to feed urban populations, security concerns regarding the border with China, and anti-urban rural socialist ideologies (see Forbes and Thrift 1984). More recently agricultural expansion has been used by socialist and non-socialist states as a means to control marginal ethnic groups living in remote highland areas, many of which are disliked or distrusted by lowland states (Dery 2000). Many of these minority groups practice a traditional form of agriculture known as shifting cultivation where they rotate fields rather than crops. This entails a cyclical process of clearing where a parcel of forested land is 'slashed and burned' before being harvesting for one or two years, and then left alone as farmers shift to the next plot and repeat the process. Depending on the size of land available to shifting cultivators the cycle may take 20–30 years to complete, allowing forests to recover before commencing the cycle again. Agricultural expansion affects shifting cultivation in two ways. First, it results in the conversion of forest land to settled agriculture thereby shrinking the 'fields' available to shifting cultivators and making their practices less sustainable because of higher rotation rates. Second, it discursively presents settled agriculture as a modern civilised form of farming in contrast to shifting cultivators whose activities are depicted as inefficient and environmentally damaging and often subsequently criminalised. In Laos, for example, the government criminalised shifting cultivation and between 1989–2000 it sought to resettle 60 per cent of its 1.5 million shifting cultivators (then 25 per cent of the country's population), or 90,000 people every year, in more conventional agricultural activities with varying degrees of success (see Evrard and Goudineau 2004). Similarly agricultural expansion in the Malaysian state of Sabah has forced people out of shifting agriculture altogether or forced them to purchase fertilisers and herbicides to retain their lifestyles, negatively impacting local ecosystems and soil fertility (Lim and Douglas 1998).

Intensifying production

A second strategy to boost agricultural incomes centres upon intensifying agricultural production. Rather than seeking new land these programmes focus on increasing yields in existing agricultural spaces. While farmers have always experimented with different crops, seeds and farming methods to boost agricultural production the most concentrated period of modern intensification began during the Green Revolution in the mid-1960s. The Green Revolution represents a classic application of modernisation theory being a top-down technical solution to concerns about rural poverty and food security. The programme was promoted by institutions such as the World Bank and the FAO and sought to increase the productivity of farm land by introducing new high yielding varieties (HYVs) of crops. In the Southeast Asian context most research focused on rice HYVs with new strains being bred at the International Rice Research Institute in the Philippines. In addition to HYVs the Green Revolution also promoted modern agricultural technologies such as pesticides, herbicides and fertilisers as well as the new farm machinery such as tractors.

The impacts of the Green Revolution have been mixed. On the one hand it has succeeded at the regional scale in its core goal of increasing overall agricultural output. Average paddy yields have increased across the region but particularly in places that use irrigation rather than rain as their main source of water as it allows multiple yields of HYVs during dry seasons. Hence the benefits have been concentrated in lowland irrigated areas of the Philippines, Indonesia and Vietnam while results have been less spectacular in other more rain-dependent areas such as Thailand. Some places have not benefited at all as their existing climatic and agricultural conditions were unsuitable to HYVs and traditional varieties of rice are still used (Hafner 2000). One key problem then has been the uneven distribution of benefits across the region. Another issue has been the increased reliance on fertilisers, pesticides and herbicides due to the greater susceptibility of foreign HYVs to local pests and climatic conditions than traditional varieties, which have more natural resistance. This has made farmers more reliant on global agribusiness corporations who sell chemical inputs and at risk of shifting currencies and prices. This caused stress during the Asian economic crisis when plummeting currencies saw agricultural supplies double or triple in some places (see Bourgeois and Gouyon 2001) and has contributed to health problems. A study by Grandstaff and Srisupan (2004) in central Thailand, for example, found that nearly all the rural pesticide sprayers they sampled had experienced some form of acute pesticide poisoning, including being 'knocked out by the drug'. Similarly poor health and safety practices and poor quality pesticides, many of which are produced by but banned in developed countries, resulted in pesticides causing 28 per cent of all reported cases of acute poisoning in Indonesia in the 1980s (Jeyaratnam 1990).

Other concerns have focused upon who has access to Green Revolution technologies. The high costs of purchasing new seeds, pesticides, herbicides and mechanised technologies make it difficult for poorer farmers to participate within new programmes without access to some sort of external financing. As a result fears have been raised that Green Revolution programmes are more likely to benefit wealthy farmers over poorer ones and exacerbate, rather than eradicate, rural inequality. This may have occurred in some places, however in other areas poor farmers have managed to access finance from unconventional sources, enabling them to benefit from the new technologies. Ghee (1989), for example, found that smallholder farmers in the Muda area of northwest Malaysia were able to access innovative low interest credit schemes that allowed them to maintain small landholdings that would have been unprofitable without the increased outputs of HYVs. Similarly Banzon-Bautista (1989) found that poor Filipino farmers who couldn't access formal credit schemes were able to borrow from relatives who had migrated abroad to stay abreast of change.

The mindset established during the Green Revolution has never really faded. While the term is not used as much today its core approach of using improved technology to intensify agricultural production still shapes the thinking of governments, international development donors, agribusinesses as well as farmers themselves. The latest and most controversial manifestation of the Green Revolution has been to genetically modify seeds, rather than breed them, in search of more productive and resistant varieties. While the stated aim of genetically modified (GM) food crops is to increase agricultural outputs many critics are concerned that the new GM seeds are oriented more towards profits for international agribusiness corporations than boosting the incomes of small farmers. Much GM research in Asia, for example, has focused on developing HYVs that are herbicide resistant, which means farmers can apply more herbicides to their crops without fear of killing them. This boosts agribusiness profits as they own the GM seed and the herbicide while potentially having deleterious effects on the health of farmers and the environment. Similarly many activists fear that lack of regulations, poorly informed populations and suppressed civil society movements make Southeast Asia an ideal place for trialling experimental GM products. In the Philippines, where civil society is relatively free (see Chapter 5), a coalition of church groups, NGOs and socialist organisations have mobilised against GM initiatives in country, voicing concerns about the unknown impacts GM products may have upon health, livelihoods and agriculture (see Box 7.2; Aerni and Rieder 2000). Nevertheless wealthier countries, such as Singapore and Malaysia, are expanding into biotechnology research through the ASEAN Committee on Science and Technology with the aim of expanding GM technologies throughout the region (see Hindmarsh 2003 for an overview of debates).

The modernist approaches of the Green Revolution have successfully boosted agricultural outputs but they have also made previously self-reliant farmers dependent upon international markets for pesticides,

Box 7.2

Biopiracy and rice modification in the Philippines

Biopiracy, also known as bioprospecting, is a term used to refer to the theft of biological knowledge from indigenous communities by private corporations, universities and governments who patent that knowledge under international law. A 'bioprospector', for example, may meet with indigenous communities to learn about their traditional healing practices and identify useful medicinal plants. A patent is then taken out, not on the plant itself, as it is impossible to patent an existing living organism, but on specific chemicals and compounds that are present within the plant. Patents can be applied to compounds that are isolated and extracted from organisms, or on a 'cultivar', a new variety of the organism, in which the properties of the original have been genetically enhanced. Under the Convention on Biological Diversity developing countries that are rich in biodiversity, but poor in biotechnology, are entitled to a share of the profits commercial organisations gain from patenting organisms derived from their resources and traditional knowledge. However contracts between corporations and national governments for bioprospecting have often been viewed as grossly unfair, if they exist at all, and it is this unethical behaviour that has given rise to the term 'biopiracy'. Very often the community that possesses the crucial medicinal or agricultural knowledge sees none of the proceeds from the subsequent international patent.

In the Philippines debate has focused around the production of GM food crops, and access (or lack thereof) to traditional biological and genetic resources and knowledge. Since the beginning of Green Revolution in the 1950s Filipino farmers have abandoned approximately 3,500 indigenous rice species for just eight 'modern' HYVs that are now grown across millions of hectares of land. Many of the earlier varieties have been lost forever while others have been stored in seedbanks for further research and possible genetic manipulation at the International Rice Research Institute (IRRI) in Manila. A further consequence of the commercialisation of Filipino farming has been a reduction in the amount of land devoted to subsistence fruit and vegetable plots, which have traditionally been an important source of vitamins. Vitamin A deficiency, which can cause blindness in small children is now common, and a major public health concern.

The response from IRRI and other global agro-corporations has been to develop a new 'golden rice' species, genetically modified to be high in beta carotene which provides Vitamin A. This has caused controversy among local NGOs for two main reasons. First, they argue that the causes of vitamin A deficiencies lie with agro-corporations and institutions such as IRRI in the first place, as they are the ones who modified local diets by promoting homogeneous rice varieties. Local NGOs believe grassroots rural development programmes oriented at overcoming rural poverty, so local communities can grow indigenous fruit and vegetables rather than intensify local dependence on foreign technologies and expanding foreign profits, are a better solution. Second, the introduction of vitamin A enriched rice is viewed as a 'Trojan horse' put in place by the agro-food corporations to make GM foods more acceptable to the public, after which it will be easier to introduce new products currently under development. Such new products are criticised for being top-down, profit-oriented and are not always appropriate or wanted. In addition 'new' genetically engineered organisms, such as golden rice, are sometimes pirated, having originated from the seedbanks that house traditional indigenous knowledge and crops.

Sources: Shiva (2000), Alier (2000)

fertilisers and seeds. While economically empowering the process has been top-down and potentially socially disempowering as local knowledge of seeds, crops, pests and sustainability has been disregarded as generic HYVs and associated chemical packages are encouraged across the region. A more inclusive approach to intensifying agricultural production has evolved from the alternative grassroots development approaches discussed in Chapter 1. Grassroots rural development programmes support and empower farmers by focusing on their needs and concerns and helping them develop their own knowledge to increase or sustain production. Seen as bottom-up, rather than top-down, such strategies are now popular among the development industry and farmers alike. The Integrated Pest Management Programme (IPMP) is an example of a grassroots approach that has been supported in Southeast Asia by the FAO since the early 1990s. The IPMP encourages farmer field schools that bring farmers together to exchange ideas and ask questions about issues they wish to know more about. The programme is valued for encouraging gradual change in farming practices while retaining the power and dignity of the farmer who controls the production of knowledge, rather than acting as a receptacle for knowledge developed in laboratories elsewhere (see Winarto 2004). Such approaches, which minimise external inputs, are leading to more productive, self-reliant, equitable and sustainable agricultural outputs for rural farming communities.

Land reform

One of the most longstanding and significant issues facing rural communities in Southeast Asia is that of land reform. Prior to colonisation most land in Southeast Asia was 'owned' according to locally recognised rules known as customary tenure. These were rules that were specific to particular areas whereby the rights to use the land were allocated according to the cultural norms of that place. Rights may be inherited according to lineage, allocated to households according to power and prestige or may be shared among a group within the community. In some areas households may have more or less exclusive rights to spatial parcels of land while in other places harvesting rights may be allocated to a social group according to species: elderly women, for example, may have the right to harvest a particular tree irrespective of where that tree is located. Such customary land tenure systems have developed over centuries, evolving with the diverse societies, cultures and environments that make up the region. Very little was written down or mapped, instead rules were embedded and reified within everyday practices and institutions and passed down and modified through local negotiation and agreement.

The lack of formal documentation in regards to customary tenure has proved to be a fundamental weakness in community negotiations for land rights during the formation of modern Southeast Asian states. Under colonialism, for example, rural spaces were vulnerable to land grabs by European powers that either did not recognise customary tenure or made the formalisation of customary tenure difficult through complicated and bureaucratic land claims processes. Any land that was considered under-utilised could be declared unused by the colonialists and appropriated for colonial control and development or gifted to indigenous 'friends' in return for their support. When land ownership was granted to local communities it was local male village leaders who were the most likely to benefit, while women and those who did not 'fit' into spatial models, such as those that may have had claims to a particular species or practice (such as collecting fallen fruit), the most likely to miss out. Powerful new maps were drawn up that secured the rights of the fortunate and new laws designed to prevent claims from the dispossessed. The new maps transformed millions of Southeast Asians from being customary landowners to land tenants, often without their knowledge or participation (see Eaton 2005 for a discussion).

The eradication of customary tenure has continued since independence with states either claiming land as their own or selling it off to private interests. This has lead to some entrenched inefficiencies in rural production systems. In the socialist states of Vietnam, Laos and Myanmar, for example, the state took control of all land and attempted to arrange farmers into rural collectives that would pool resources to work communal land. However the lack of individual profit-making opportunities acted as a disincentive for farmers and agricultural production was poor as a result. In market-led states large private landowners have evolved, alongside smaller farmers and a rural landless labouring class. This has led to a variety of largely inefficient landowner–labourer relations that include:

- landless labourers being paid a wage to work landowners' land much like someone would get paid to work in a factory;
- sharecropping arrangements where labourers live and work on the land in return for an agreed proportion of the profits;
- leasing arrangements where the labourer is granted land to lease for his or her own subsistence use in return for working the landowners' fields for an agreed period of time.

All three approaches can be criticised from economic, political and social justice perspectives (see Borras and Franco 2005). Economically the systems are inefficient as the tenant or labourer does not have much incentive or capacity to boost the productivity of the land. Politically such arrangements are unpopular and have led to peasant movements lobbying for political change. Socially such arrangements are likely to increase inequalities rather than lessen them, contributing to poverty for landless labourers and great profits for landowners. The inefficiencies of both socialist and market-led land ownership models have led to a number of initiatives to reallocate land more fairly.

In Indonesia, for example, the decentralisation processes described in Chapter 4 are contributing to a greater recognition of customary tenure arrangements. Local *adat* (customary) community structures managed to survive the Suharto years by handing over formal power to state authorities but maintaining important community functions (McCarthy 2005). However they lost the right to much of their land, including access to forests that totalled 74 per cent of Indonesia's terrestrial land area, when Suharto came to power. In non-forested areas the land titling process was slow and inefficient, enmeshed in claims of corruption, inaccuracy and incompetence, with land titles only granted to 30 per cent of all agricultural lands (see Thorburn 2004). This has contributed to poor agricultural production rates as well as confusion, intimidation and violence as communities have sought to protect their rights in an uncertain legal environment. Since decentralisation processes have been adopted in post-Suharto Indonesia local authorities have gained responsibility for allocating land titles. While this can contribute to local scale corruption it has also lead to a revitalisation of *adat* customary land claims. In Bali, for example, local NGOs, *adat* leaders and local government have worked together to map out community scale development directions involving land ownership, use and community controlled resources with empowering results (Warren 2005). Decentralisation is assisting communities to formalise customary tenure systems.

While changes in land allocation in Indonesia are an outcome of broader decentralisation approaches, specific land reform programmes have been introduced in Southeast Asia for political or economic reasons. The Philippines, for example, has the most unequal distribution of land in the region, partly a result of land allocation policies under the Spanish, with a small group of elite landowners inheriting huge estates that dominate agricultural production throughout the country. As a consequence the agriculture sector employs over 40 per cent of the country's workforce but only attracts 19 per cent of GDP, with the rural poor comprising two-thirds of the total poor population in 1999 (Borras 2001). Economic inefficiencies in the rural sector, alongside social movements demanding change, has led to a series of Filipino land reform programmes, including the most recent Comprehensive Agrarian Reform Programme (CARP), which was introduced in 1988. The CARP programme has made some impressive achievements including the redistribution of over 42 per cent of total agricultural land by 2000, contributing to a lessening of inequality within rural spaces (see Borras 2001). However the programme has also attracted criticism for its complexities, loopholes and unintended impacts, which include the eviction of peasants from their plots. However, more encouragingly, it has also led to a radicalisation and politicisation of peasant identities (see Box 7.3).

Box 7.3

Land reform conflicts in the Philippines

The Filipino agricultural system has been characterised by feudal relationships since colonialism when influential families secured large estates that are now worked by landless peasants. This arrangement is inequitable and a source of entrenched rural poverty as landless households are poorly paid, are economically dependent on the landowner, and possess few resources to improve their situation. There have been repeated attempts at rural land reform, most of which elite families have managed to avoid, however the most recent attempt, CARP, initiated by President Aquino in 1988, appears to be making some headway.

CARP essentially requires landowners to sell the majority of their agricultural land to the government, who, in turn, sells it to peasant workers through a government assisted loan. Proponents of CARP argue that the programme will jump-start rural development and reduce rural poverty as elite landowners will have more capital to invest in industrial projects and non-farm industries while peasant labourers will have greater incentives to invest in the land and maximise productivity and innovation. There has, however, been considerable landowner resistance. This has included evicting peasants from land before CARP processes take place so they cannot be considered beneficiaries, reclassifying land to avoid redistribution, using their local power and wealth to influence or corrupt government land redistribution decisions, as well as threatening peasants who become involved in CARP processes.

CARP has performed particularly badly on the sugar growing island of Negros where landowners who make up 3.9 per cent of the population own 72 per cent of rural land, leaving 82 per cent of rural Negranese landless (see Figure 7.3). Landowner resistance to reform has been strong with some choosing to illegally evict peasants from their land before claims can be made. Civil society organisations have responded by attempting to empower peasant beneficiaries by guiding them through legal systems and informing them of their rights, protecting them from regressive landowner tactics, and encouraging them to form farming collectives once they gain legal ownership of agricultural land.

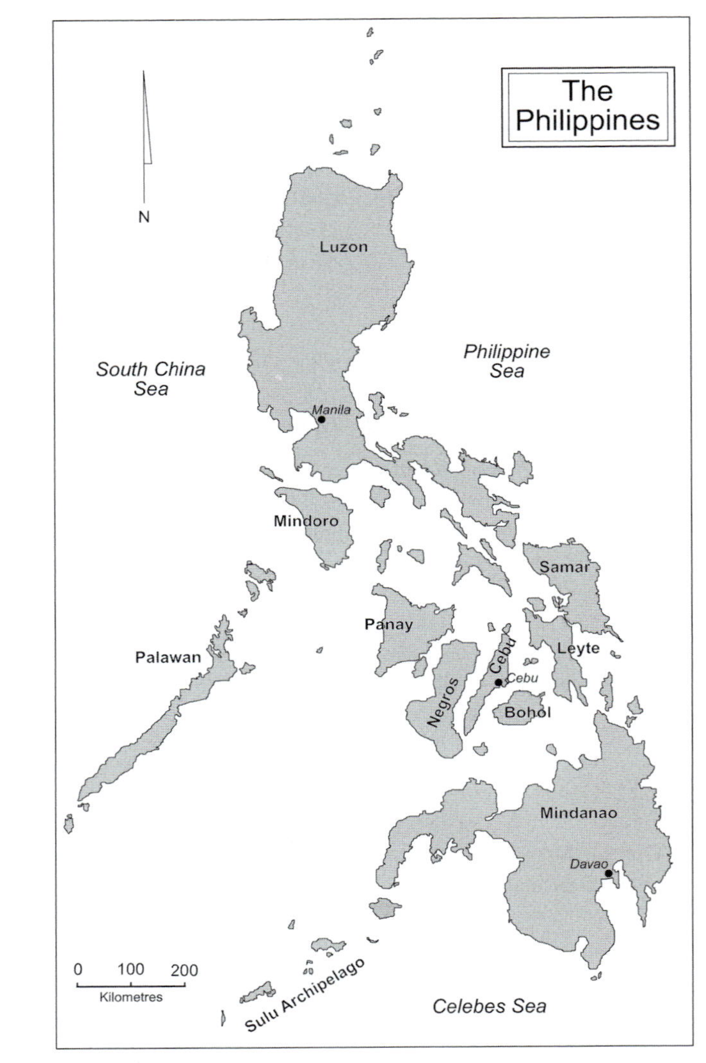

Figure 7.3 ***The Philippines.***

While NGOs have undertaken a range of conventional resistance activities, such as marches and land occupations, they have also implemented radical education campaigns oriented at transforming peasant imaginaries so they see themselves as 'rightful' owners of the land. The education programmes attempt to move peasants beyond the existing 'culture of patronage and indebtedness' they feel for landowners. One NGO worker explained it as follows:

> Farmers may go to the landowner in the off season asking for food. The landowner will 'gift' them rice (which is not really a gift because it was probably deducted from their pay or denied through not paying

> benefits). But nevertheless the workers will be grateful and loyal to landowners because it is ingrained in their culture and way of thinking.
> (Diprose and McGregor forthcoming)

To empower peasants and change this culture NGOs present an alternative Filipino history that repositions peasant beneficiaries (and their ancestors) not as grateful tenants, but as exploited labour. The new histories stress the injustice of current and past relationships and outline the role of peasants in previous rebellions against injustice. As another NGO worker remarked, 'we want movements within . . . minds – from uneducated peasant to educated, conscious, organised, resistant peasant'. Such narratives have helped bind geographically, politically and economically marginalised and isolated peasants together under a collective resistance identity that is empowering their attempts to access the land rights promised to them under the CARP legislation.

Gradon Diprose, Department of Geography, University of Otago

Sources and further reading: Diprose and McGregor (forthcoming)

The socialist states of Vietnam and Laos have also undergone land reform as part of their moves towards greater economic liberalisation. The focus of this reform has not been reallocating land from private interests but reallocating land from rural collectives to individual farming households. In both countries the state retains ownership of the land but leaseholds are provided to farmers, which gives them some security and incentive to invest in their lands and enhance production. In Vietnam the reform processes began in 1988 when the central authority devolved responsibility for allocating land to commune scale authorities, and was largely completed by 1990. Farmers can lease state land for up to 50 years and have considerable control over that land, including the ability to rent it out to other farmers. Despite fears, and some well-recognised cases of corruption at the local scale (see Kolko 1997), the land reform process is generally recognised as having been remarkably successful in redistributing land fairly and equally (Griffin *et al.* 2002). The success of the Vietnamese programme reflects the principles of equality that originally underpinned the formation of the socialist state, close monitoring by local communities among whom commune leaders live, as well as monitoring by central government that would publicise breaches in the state-owned media (Ravallion and van de Walle 2004). Laos is undergoing a similar process funded by the World Bank in which farmers practising settled agriculture receive land titles and security to state-owned land while shifting agriculture on forested land is illegalised. The programme has not been as successful in application as the Vietnam case leading to the rather dour conclusion by Ducourtieux *et al.* (2005: 521) that 'land allocation is starting to be perceived as a policy for eradicating shifting cultivation and not as a tool to fight poverty'.

De-agrarianisation

While rural development schemes have traditionally focused on agrarian development there is a growing body of work which suggests non-farm activities are becoming as important, or possibly more important, in sustaining rural livelihoods. The shift away from agricultural pursuits to non-farm sources of income and employment is known as the process of 'de-agrarianisation' (see Rigg 2005 for a discussion). This can be a full-time shift where households sell their farmland and seek alternative income generating activities or a part-time shift by which off-farm income earning opportunities supplement household agricultural returns. The shift towards de-agrarianisation reflects increasing land pressures, changing expectations among rural youth, as well as an expansion in alternative rural income earning opportunities. Land pressures have resulted from rapid population growth in rural spaces resulting in less arable land available per household despite agricultural expansion programmes. With only small returns possible from small plots many farmers have chosen to sell their land, often to large landowners who pay high prices in order to set up industrial farming systems. Changing expectations reflects improved education and communications technologies in rural spaces that is changing the 'software' or psychology of rural communities, particularly youth, exposing them to alternative lifestyle opportunities and non-farm existences (Rigg 2005). In other words people are not necessarily being forced off farms due to adverse economic conditions but are choosing non-farm livelihoods as more desirable lifestyle options. This is only possible because of an expansion in non-farm employment options that reflects diversification of rural economies, greater rural mobility, improved educational standards and a trend towards rural industrialisation, particularly along transport corridors and in the fringe areas of EMRs. Non-farm employment opportunities are not equally distributed throughout Southeast Asia and as a result de-agrarianisation is likely to be most advanced in wealthier or more developed rural spaces located close to roads and/or larger urban centres. Four aspects relating to the diversification of rural economies will now be discussed.

Small scale business opportunities

The shift from subsistence to commercial agriculture has seen the evolution of cash economies throughout rural spaces of Southeast Asia. Rural households have more capacity to purchase and consume goods enabling a range of new income earning opportunities. Many of these take the form of locally-owned small businesses that involve such activities as:

- the trade of goods into and within rural spaces through shops and petty traders;
- the provision of services and trades such as transport, repair, construction and general labouring;
- the hiring out of equipment, such as tractors and tools for agriculture;
- cottage or home-based industries that produce handicrafts and other goods;
- food production and sales through food-stall operators.

While some of these opportunities require relatively little skill or start up capital, and hence have small returns, other activities, such as equipment hire and provision of transport vehicles require substantial investment and can have high returns. Rutten (2003), for example, has discussed the evolution of a rural capitalist class in Malaysia and Indonesia, who control valuable assets such as combine harvesters which they hire out to farmers in return for large profits. The success of this class challenges long-held assumption that land ownership is a prerequisite for rural wealth and success.

Industrial manufacturing opportunities

A range of industrial manufacturing industries that have traditionally been located in urban centres are being relocated to rural spaces. While socialist economies have long sought to decentre industrial activities the trend is more recent in market-led economies. Today a combination of factors including government policies oriented at slowing rural–urban migration, improved transport, communication and educational infrastructure in rural spaces, and cheaper land and labour costs are contributing to rural industrialisation. Many of these industries are related to the processing of agricultural products such as refineries, canneries and packaging plants, as well as pulp and paper mills for wood products. Other industries, such as footwear or textile industries, may have little to do with agriculture but are relocating to rural spaces in response to government incentives and market forces. Rural factory work may be considered very desirable as employees can work with friends, for relatively good pay, in positions that are not as physically demanding as farm work, and do not require people to migrate away from home to find employment. Youth in particular appear to be attracted to rural factory work signifying long-term changes in the economic composition of rural spaces.

Mobile income earning opportunities

Another important component of de-agrarianisation is the increased mobility of rural people that has been enabled by improved transport infrastructure. As mentioned in Chapter 6 many people from rural spaces migrate to urban spaces in search of employment as well as services and new lifestyle opportunities. Some of these people, particularly if the whole household moves, are lost to rural spaces and only have minimal contact with their non-urban roots. Others however, particularly those migrants that leave behind members of their rural household, continue to play a very important role in rural economies by sending back a proportion of their pay as economic remittances. As these migrants are economically influential and are likely to return periodically or permanently it is changing the way researchers conceptualise rural households. They are no longer seen as spatially contained units but instead conceptualised as centres of diverse networks that can extend into urban spaces and across territorial borders. Jones and Pardthaisong (1999), for example, found in their survey of rural villages in Thailand that over 80 per cent of households had at least one member travel internationally to pursue work opportunities. While spatially distant these migrants are important income earners for rural households and are freeing up farmers from traditional agricultural pursuits.

Tourism industry opportunities

In some rural areas an important example of non-farm rural employment derives from the tourism industry. Southeast Asia is now well established as a tourist destination with people coming from across the world to experience the cultures and landscapes of the region. While Singapore and Bangkok are popular with business travellers and shoppers, most recreational tourists leave the big cities to seek out sun, sand, sea and adventure in rural spaces. Some popular rural tourism destinations include parts of Bali and Tana-Toraja in

Indonesia; the southern Thai islands around Khao Samui and Phuket, as well as the rural areas surrounding Chiang Mai in the north; Boracay island in the Philippines; and the Angkor Wat areas around Siem Reap in Cambodia. In addition to these key destinations is a well-established and expanding tourist infrastructure that stretches across the region, including Myanmar, which is permanently inundated by both short-term tourists and long-term backpackers. For rural communities the spread of tourism to their localities greatly enhances the prospects of non-farm employment. Households can engage in homestays or invest in small scale hotels and restaurants while people can find or create jobs in hospitality industries such as hotels, bars, restaurants and travel agencies. For relatively low outlay small household businesses can earn large profits by engaging in niche industries such as massage, transport or travel guide services. In addition new markets arise for cultural products and performances boosting sales of handicrafts, artwork and souvenirs as well as dance and musical performances. As a consequence most Southeast Asian governments give tourism a high priority within their national economic development plans.

Conclusions

There can be no doubt that rural spaces lag behind urban spaces in terms of their economic and social development. While the division is not as pronounced in non-capitalist economies urban centres still boast better incomes, less poverty and better services than in rural areas. Perhaps this is inevitable to a certain extent given the lower population densities of rural spaces and the difficulties in providing adequate and easily accessible services to all. There are some positive trends, however, of which the diversification of rural incomes is one of the most important. Preston (1989) observed in his study of two villages in central Java that people were becoming 'too busy to farm' given the range of alternative income opportunities arising within the villages, as well as the increased mobility of those seeking work elsewhere. These processes are providing a conduit through which the benefits of urban development, including wealth and enhanced employment opportunities, can be redistributed to households in rural spaces. While this will never overcome rural/urban inequalities it can help narrow the divide and raises challenging questions for development planners. It is possible, as Rigg (2005) argues for example, that rural development policies should focus less on agrarian production, as they have in the past, and more on improving non-farm employment opportunities as the key means of boosting rural economies. It is in non-farm employment that rural spaces are experiencing the greatest degree of growth and there is great potential here to overcome rural inequality, as traditional distinctions between landowners and the landless are lessened.

This does not mean the agrarian sector should be neglected although the trends there, in terms of overcoming inequality, are not as promising. The agrarian sector is becoming increasingly commercialised and dependent upon TNCs for agricultural inputs and new technologies while global markets are dictating returns on cash crops. These conditions favour large industrial agricultural plantations that benefit from economies of scale when compared to smaller landowners. While some small farms have prospered as a result of agricultural intensification the option of selling land to industrial plantations is becoming increasingly appealing given the high prices being offered and the alternative income opportunities available. Indeed migration-based agricultural expansion programmes no longer give land to settlers, instead they become shareholders and labourers of private economic enterprises. Rural development programmes that centre on enhancing agricultural production may well lead to greater inequality by benefiting large landowners over landless labourers or those that have moved out of the agricultural sector. However the industrialisation of agriculture is not uniform across the region and small farms are still the norm in many areas. In addition some of the more progressive land reform programmes in the region seem to making headway in addressing core land ownership problems that contribute to rural inequality. This would appear to be the case in the Philippines and Vietnam where peasants are securing land tenure and thereby expanding the rural development opportunities available to them. Given the diversity of challenges and trends in rural spaces development policies need to be sensitive to the particularities of place when designing policies oriented at overcoming inequality and maximising rural livelihood opportunities.

Summary

- **Rural spaces are being fundamentally transformed by development processes but are attracting a diminishing proportion of national wealth and services.**

- **Most rural spaces have undergone the transition from subsistence to commercial agriculture and there is a more recent trend towards large industrial agricultural estates.**
- **Agricultural expansion programmes have successfully cleared new land for the rural poor but have come under criticism for their recent shift towards shareholding rather than landowning, the impacts they are having on forests, and their role in nation-building exercises.**
- **The controversial top-down modernisation approaches of the Green Revolution continue today through genetic modification; in contrast grassroots NGOs are focusing upon bottom-up community-based knowledge.**
- **Land distribution is extremely unequal in market-led economies and is being addressed through land reform and decentralisation programmes.**
- **De-agrarianisation is opening up new non-farm employment opportunities in rural spaces across the region and is posing challenges for authorities who have traditionally focused their development efforts upon the agrarian sector.**

Discussion questions

1. Compare the ways rural households are being empowered and/or disempowered by rural development.
2. Discuss how and why grassroots and modernist development theories approach rural development in different ways.
3. Explain the concept of de-agrarianisation, its multiple dimensions and its importance to rural spaces.
4. Compare the pros and cons of investing development efforts in agrarian and non-agrarian sectors to boost rural quality of life.

Further reading

For an excellent succinct introduction to de-agrarianisation debates see:

Rigg, J. (2005) Poverty and livelihoods after full-time farming: a Southeast Asian view. *Asia-Pacific Viewpoint* 46(2): 173–184.

For more in-depth case studies on rural transformations see:

Rutton, M. (2003) *Rural Capitalists of Asia: A Comparative Analysis on India, Indonesia, and Malaysia*. London: Routledge.

Hart, G., Turton, A. and White, B. (1989) *Agrarian Transformations: Local Processes and the State in Southeast Asia*. Berkeley: University of California Press.

Useful websites

The key United Nations institution for rural development is the Food and Agriculture Organisations at www.fao.org

An interesting regional website for GM/Green Revolution technologies is the Manila-based IRRI at www.irri.org

Many governments provide services on agrarian reform. See the Philippines www.dar.gov.ph and Malaysia www.felda.net.my/felda/default

8 Transforming natural spaces

Introduction

Natural spaces are defined in this book as spaces in which naturally occurring species, ecosystems, processes and phenomena dominate the physical landscape or characteristics of that space. This does not mean humans are absent in such spaces or that nature exists in some pristine or untouched condition, indeed apart from deep beneath the earth's surface it is possible that no such spaces continue to exist. Instead human influence is present but is less dominant than in rural or urban spaces where landscapes are epitomised by rows of buildings or agricultural fields. Hence natural spaces in Southeast Asia include seas, rivers, forests, the atmosphere, soils and subsurface processes, all of which are shaped by natural processes but also bear the imprint of human actors. Humans have long fished from the seas and rivers, hunted from forests, released gases into the atmosphere and sprayed chemicals onto the soils. Other terms people use to refer to such phenomena include 'natural resources', which is rejected here for concentrating upon only those aspects of nature that have some economic value or human use; and 'the environment', which is a more general term that can refer to human and non-human surroundings. This chapter explores the role of natural spaces within Southeast Asian development in three main sections. The first looks at how natural spaces are conceptualised in development theory and applied to the Southeast Asian context; the second reviews the pressures and conflicts surrounding three types of natural spaces; and the third section looks at the risks and dangers, some of which became horribly apparent during the Asian tsunami of 2004, natural spaces pose to human development.

Nature and development

Different development theories value natural spaces in different ways. Modernist, neo-Marxist and neo-liberal theories are anthropocentric or human-centred theories that value natural spaces for what they can contribute to national scale economic development. In general these theories are interested in the transformative economic value of nature, in other words they seek to establish industries that convert natural spaces into goods that can be used or sold to propel economic development. Hence forests are valued for timber; rivers are valued for the energy that can by produced when dams are built; seas are valued for their catch. Grassroots and alternative development theories are also anthropocentric but are more likely to be interested in the existence values of nature, or the value that nature has to humans as it now exists. These theorists concentrate upon how people interact with natural spaces at the local scale and are more likely to value forests, rivers and seas for the livelihoods such spaces currently provide for locally dependent communities. A third perspective comes from environmental NGOs such as Greenpeace and World Wildlife Fund that are becoming increasingly influential in development decisions regarding natural spaces. These environmental groups share many of the anthropocentric concerns of grassroots agencies but also harbour ecocentric or ecologically centred values in which nature has intrinsic value, which means it is valued for its own sake, irrespective of its usefulness to humans. All three perspectives have attracted supporters at local, national and international scales making any decisions about the management and development of natural spaces inherently contested and controversial.

The catch-all solution to these tensions has been the adoption of the concept of sustainable development that has been popular at the global scale since the Rio Earth Summit of 1992. Sustainable development promotes development initiatives that meet the needs of the existing generation without comprising the opportunities available for future generations. As such attention is focused upon the sustainability or longevity of current interactions with natural spaces with a view to ensuring natural spaces are not irreversibly damaged and that ecosystems and processes are able to replenish themselves. This is achieved through a range of means that include improving dialogue and understanding between stakeholders at local, national and international scales. Southeast Asian countries have shown considerable interest in the principles of sustainable development, with the ASEAN Vision 2020 calling for a 'clean and green ASEAN [by 2020] with fully established mechanisms for sustainable development to ensure the protection of the region's environment, the sustainability of its natural resources and the high quality of life for its peoples' (ASEAN Secretariat 2002: 2, cited in Elliot 2004: 188–189). There is an ASEAN Ministerial Meeting on the Environment every three years and every country now has a ministry or upper level executive agency responsible for sustainable environmental management. Despite this encouraging progress serious criticisms can still be raised about the sustainability and equity of current approaches to natural spaces. The region's forests are rapidly declining with deforestation rates in Indonesia among the highest in the world; rivers and seas are being polluted and over-fished raising questions about their long-term viability; atmospheric pollution often drops below World Health Organisation standards due to urban emissions and forest burning for agriculture; and soil profiles are being degraded through overcropping and the introduction of chemical fertilisers and pesticides. Despite official support for the principles of sustainable development natural spaces in Southeast Asia are experiencing serious and ongoing degradation.

There are many reasons as to why this is the case. One of the most important is the priority given to economic development strategies at the national scale. While attitudes are broadening there is still an overwhelming focus on raising GDP and the exploitation of resources within natural spaces is one means of doing this. Governments throughout the region have profited from selling natural resources to international and national logging, fishing or mining interests to increase GDP and promote national development. A second reason relates to the growing wealth and associated consumption patterns of rapidly increasing urban populations. Wealthier populations consume more resources, which puts pressure on natural spaces where food, timber and energy sources are often located. A third factor relates to FDI-oriented development, which puts pressure on governments to relax environmental standards in order to create an attractive and inexpensive investment environment. This enhances the likelihood of environmental accidents and pollution that damage natural spaces.

Other factors include a lack of funding for environmental agencies to develop policies, monitor practices and police offenders as well as a lack of integration of environmental principles into other sectors such as transport or manufacturing (see Elliot 2004 for a discussion). Consequently despite official acceptance of the principles of sustainable development the clashes between local, national and international interests continue to take place. Clashes in three types of natural spaces will now be reviewed.

Pressures and conflicts in three types of natural spaces

Forested spaces

The forests of Southeast Asia are among the most diverse of any in the world. The region is relatively well forested with forests covering 50 per cent of the land area, compared with a global average of only 30 per cent, however these figures include degraded forest land and plantations (Elliot 2004). The rate of forest loss for the region is among the highest in the world at 1.3 per cent per annum, compared with a global average of just 0.18 per cent, and biologically rich old growth or primary forest is being lost at rates exceeding 2 per cent per annum. These figures equate to an annual average loss of 2.8 million hectares across the region between 2000–2005, with Indonesia experiencing the greatest overall loss of 1.9 million hectares per annum followed by Myanmar, Cambodia, the Philippines and Malaysia (see FAO 2007). Only Vietnam can boast increasing forest cover although some of this is in the form of forest plantations rather than an expansion of natural forests. At immediate risk from rampant deforestation are the endemic species and ecosystems that exist in the forested spaces of Southeast Asia, many of which remain unnamed and undiscovered. Forests are rich in plant, insect and bird life, many of which are on the verge of extinction, however it is the large land mammals that have captured the imagination of conservationists worldwide. Some better known mammals include Indonesian and Malaysian orangutans, Sumatran tigers and rhinoceroses, the Javanese rhinoceros, the Malaysian tapir, the Asian elephant as well as a range of gibbons, monkeys and other vertebrates. Many of these mammals are now on the endangered list; less than 500 Sumatran tigers are estimated to exist and only around 60 Javanese rhinoceros, making them possibly the rarest large mammal living in the wild today (Linkie *et al.* 2003; Fernando *et al.* 2006). These two species were once found throughout Southeast Asia but are now restricted to isolated colonies in hard to access areas of Sumatra, Java and Borneo, and a recently discovered small colony of possibly five Javanese rhinoceros in Vietnam (Fernando *et al.* 2006). While orangutan numbers are not quite as low their habitats are being rapidly cleared for agricultural plantations, particularly oil palm plantation, inspiring the formation of protective shelters in Sumatra and Kalimantan. These and other threatened species have their interests represented by ecocentric environmental stakeholders both within and outside the region.

Forests also provide homes and resources for millions of people who live within or beside forests throughout the region. Most forest dwellers come from ethnic minority groups, such as the Hmong, Karen or Penan, who have retained much of their unique cultural identities by continuing to practise their traditional livelihoods, such as shifting cultivation and hunting and gathering, in remote forested areas. These livelihoods are threatened by forest loss as their activities become less productive and sustainable in smaller forested spaces. However it is not only forest dwellers who rely on rainforests for their livelihoods but also those neighbouring rural settlements who supplement rural incomes with forest-based pursuits in which food, wood and medicines are sourced from forest spaces. The important role of forests in these people's lives is becoming increasingly recognised and integrated into poverty reduction and development strategies (see Sunderlin *et al.* 2005; Sunderlin 2006). The World Bank, for example, has written that forests are central to the livelihoods of 10 million of the poorest 36 million Indonesians (World Bank 2006). For these stakeholders forested spaces are vital to sustain the lifestyles to which they are accustomed.

Stakeholders at the state scale are less likely to be enthusiastic about the existence values of forests. Since the colonial period there have been large scale programmes oriented at cutting back forests for timber or to clear land for agricultural expansion. Many forestry departments trace their roots to colonial governments that profited from the lucrative sale of tropical woods, such as highly valued teak wood from Burma, to Europe (see Bryant 1997). To these foreign eyes forests were obstacles to development as they prevented the spread of agriculture and were seen as fundamentally unproductive. There was little awareness or concern for those communities who were reliant upon forests for their livelihoods, or for indigenous animal species that were much more likely to be hunted than protected. This colonial mindset has been inherited by independent governments who also see forests primarily as resources for, or barriers to, economic development (see Roth 2004). Such perspectives have been encouraged by institutions working within modernist development paradigms with the FAO commissioning a study in the 1960s into the 'problem' of why 'More than half the world's forests are in the tropics and yet only ten percent of the world's timber demands are supplied from these areas.' The report recommended that 'some big, well-established company . . . clear-cut the jungle and plant tree species adapted to pulp requirements; in other words, follow the example of New Zealand . . .' (PPI 1967, cited in Sonnenfeld 1998: 95). The timber harvesting industry is worth over US$10 billion per annum to the region today (FAO 2007). This doesn't include the profits from thousands of small scale illegal harvesting operations, which states lack the resources to effectively police.

Deforestation is also being driven by the types of agricultural expansion programmes discussed in Chapter 7. Within such programmes forests are seen as barriers to development, providing physical obstacles to the instigation of more productive land uses, such as settled agriculture. Consequently vast tracts of forests have been lost to allow for the establishment of plantation crops such as sugar, coffee, rubber, and more recently oil palm, all of which can contribute much more to national economies than locally significant practices such as shifting agriculture and small scale hunting and gathering. Illegal clearance operations are also increasingly common and influential as local actors and commercial institutions seek to expand their agricultural land to enhance economic returns. In Indonesia the cheapest, most common and most devastating way of illegally clearing forests for plantations is to light forest fires during the dry season that then burn large tracts of land, often much more than what was intended by the instigators. Every year local ecosystems in Sumatra and Kalimantan are shattered during the dry season as wild fires burn out of control, sending suffocating smoke hazes and associated health risks to upwind communities and countries.

The rapid rate of deforestation within Southeast Asia, the highest rate in the world according to Achard *et al.* (2002), reflects the primacy of state interests in determining the management of the region's forests. Agriculture and timber industries provide relatively accessible and important components of national incomes, particularly in less developed economies. Some hotspots of forest loss in the region include parts of Indonesia's Sumatra and Kalimantan islands, Malaysia's Sarawak and Sabah provinces (see map at Figure 4.1), as well as parts of Laos, Cambodia and Myanmar, which have experienced rapid rates of forest loss since opening their economies to foreign investors (see Bottomley 2000 on Cambodia). The existence values forests have for local communities or plants and animals are of much less importance to governments oriented at propelling their countries forwards. Indeed within many countries of Southeast Asia forests often have negative rather than positive associations, which reflect past histories of fear and violence. Communist and ethnic insurgencies have traditionally been launched from forests making them spaces to be feared by the general population rather than treasured (see Sioh 1998). Forest dwellers today are still treated with suspicion by government authorities who see them as threats to the establishment of homogeneous national identities and whose 'traditional' lifestyles don't fit modernist visions of society. Consequently forestry is seen as improving and developing, rather than destroying, the lives of forest dwellers as is evident in the following quote from the Indonesian Ministry of Forestry:

> The logging industry is a champion of sorts. It opens up inaccessible areas to development; it employs people; it *evolves* whole communities; it supports related industries . . . It creates the necessary conditions for social and economic development. Without forest concessions most of the Outer Islands would still be *undeveloped*.
>
> (FAO/GOI 1990, cited in Gellert 2005: 1351; emphasis added)

The Malaysian prime minister has referred to the Penan forest dwellers of Sarawak in a similar way:

> If you look at their plight, you will understand what will happen to the Penan if they are kept in the forests . . . to eat monkeys and maggots and caterpillars . . . We believe the Penan are humans like anyone else . . . So what's wrong with us . . . feeding them the kind of culture that will make them like any of us.
>
> (*New Straits Times*, 8 October 1991, cited in Bending 2001; see also Bending 2006: 6–7)

Such statements reflect classic modernisation approaches to traditional culture, which is seen as something to be erased and replaced by modern forms. Such discourses legitimise and popularise the cultural violence forest dwellers incur as a result of national scale development, empowering national interests over local ones.

More recently, however, growing global and national environmental concern and a greater appreciation of grassroots development theories have seen a range of more protective mechanisms be adopted. Lobbying by local and international environmental organisations has contributed to 20 per cent of the region's tropical forests be officially declared biological conservation zones (FAO 2007). These zones have immediate value for local ecosystems and threatened species and may have future economic value in terms of international carbon trading schemes and the protection of potentially valuable organisms. Their ramifications for human forest dwellers are less clear. On the one hand such protections prevent their forest lands from being cleared for timber harvesting or agriculture; however the same regulations can also be used to criminalise their traditional activities within those spaces. One of the reasons why governments have shown enthusiasm for establishing protected areas is because it brings them environmental legitimacy and power in areas where their authority has traditionally been resisted by marginal groups. Consequently the state uses conservation ideologies based on Western conceptions of wilderness, which values forests devoid of human influence (see Cronon 1996), to legitimise coercive tactics against forest dwellers and other users. In Thailand, for example, forest protection laws extending back to the 1930s prohibit clearing and burning of forested spaces as well as the use of any forest products without official permits, effectively criminalising the lifestyles of ethnic minority forest dwellers (Vandergeest and Peluso 1995). While states have not had the power to police these laws their authority is growing and large resettlement programmes, such as the Lao programme to resettle over 800,000 shifting cultivators, are being introduced. Similarly the establishment of World Heritage Areas in Thailand have seen an increase in state violence oriented at moving minority groups into more modern lifestyles (see Box 8.1). In Indonesia laws orientated at protecting production forests, mostly timber plantations, have led to the establishment of forest police that 'secure and guard the rights of the state to the forest land and products', from neighbouring local communities who have traditionally used forests to supplement their lifestyles (Peluso 1993). In other words people can be forced off their lands, not only by logging, but also by conservation (see Roth 2004).

A further concern about forest protection derives from the different policies and approaches of different states. In Thailand, for example, a strong middle class, a relatively free civil society and some innovative forms of Buddhist ecological activism have contributed to a

Box 8.1

Conserving forests not people: the Thung Yai Naresuan Wildlife Sanctuary

The Karen people living in the highlands northwest of Bangkok have had a long and difficult relationship with the Thai state. Colonial borders split their homelands in two; some living within Myanmar where they have waged a long battle with the lowland military state, and the others residing in the remote forested hills of Thailand. While modernisation and development have taken place in the lowland areas many highland people have chosen to retain their ethnic identities and traditional lifestyles as much as possible. This involves, among other things, retaining their practices of shifting agriculture where they slash and burn parts of the forest before harvesting the land for a few years and moving on to another area (see Chapter 7).

The state has had a long suspicion of these and thousands of other hilltribe people, collectively known as *chao khao*. The *chao khao* have been accused of backwardness, troublemaking, growing opium and assisting communists living in the hills during the Cold War. Their ethnicity, forested lifestyles and 'wildness' threatens the homogeneous national Thai identity promoted by the state and many, despite never having left Thailand, do not have full citizenship rights. Instead of accepting such diversity Thai authorities have interfered in Karen lifestyles in an effort to have them move into lowland areas, adopt settled agriculture, and become 'more Thai'. In the 1960s Thai authorities used modernisation theories to criminalise Karen agricultural practices, portraying shifting cultivation as a destructive and wasteful use of valuable forest resources. More recently Thai authorities have borrowed from global conservation discourses to delegitimise Karen lifestyles, claiming shifting agriculture is unsustainable and destroying the natural biodiversity values of the forest.

The recent decision to create a World Heritage Site at the Thung Yai Naresuan Wildlife Sanctuary demonstrates how environmental discourses can be used against Karen communities. International biologists converged on the area and declared its worthiness, commenting on the outstanding biodiversity and 'undisturbed' nature. The Karen people, who had lived among this forest biodiversity for centuries, were identified as a threat to the sanctuary and have since become the target of a concerted state campaign to evict them through coercion and violence:

> On April 13 in 1999, the Director General himself flew into the wildlife sanctuary, landing with his helicopter at the place where the Karen had just started to celebrate an important annual religious festival supposed to last for three days. The Director General requested to stop the ceremonies. Soon after, soldiers burned down

> religious shrines of the Karen. From April 18 to May 12, soldiers and forest rangers went to the Karen villages, demanded to stop growing rice, demolished huts and personal belongings, and burnt down a rice barn.
>
> (Buergin 2003: 386–387)

While the establishment of a World Heritage Site in Thailand is a remarkable environmental achievement it has been accompanied by violence and human rights abuses. Given that the revered ecology of Thung Yai evolved under the management of Karen communities, moving them on to protect their homelands appears particularly unjust. The Western concept of an untouched natural wilderness does not make sense in the highlands of Southeast Asia, yet in translating it to the Thai context, the state has gained a tool to oppress marginalised people.

Sources and further reading: Buergin (2003), Roth (2004), McKinnon (2005)

comparatively high level of awareness of environmental issues among the general public. Domestic environmental activists have been active since at least the 1980s and became particularly influential after floods in the south of Thailand led to the loss of up to 1,000 lives and caused US$400 million in damages in 1988. The flood was widely linked to the deforestation of the surrounding watersheds, strengthening a chorus of voices opposing deforestation in the country. In 1989 the Thai government acknowledged these critics by making the landmark decision to ban any further logging of primary forests within the country, although by then most of the easily accessible forests had already been harvested. Alongside the environmental movement have been the actions of Buddhist monks who have been busy establishing forest temples that can reach up to 500 hectares and protect the livelihoods and ecosystems of human and non-human communities within (Wester and Yongvanit 2005). The protection of Thai forests has not been replicated to the same extent in other parts of the region and, in some ways, has exacerbated deforestation rates within neighbouring territories. Faced with a logging ban, a swelling environmental consciousness, but continuing domestic demands for wood products, Thai companies have crossed into the borders of Cambodia, Laos and Myanmar to source cheap timber there. A 1998 logging ban enacted in China has had a similar impact on Southeast Asian forests, increasing the demand and value of the region's wood products (Lang and Chan 2006). A regional hierarchy of environmental protections is evolving that is resulting in the transfer of environmentally damaging industries from more protected, and often more economically developed, countries to their less protected neighbours.

One solution to the divisive policies surrounding the harvesting and protection of forested spaces has been to involve local communities and decentralise decision-making regarding the management of natural spaces. This approach, known as social or community forestry, is influenced by grassroots development theories and seeks local participation and empowerment, broader recognition of the multiple interests and values forests hold for different stakeholders, long-term sustainable initiatives, and a fairer distribution of benefits. As such community forestry fits within the broad parameters of equitable development and is increasingly supported by international development institutions and through the efforts of the Bangkok-based Regional Community Forestry Training Centre for Asia and the Pacific (RECOFT). Community forestry is well advanced in the Philippines and Vietnam where 37 per cent and 12 per cent of forested land respectively is officially managed under community forestry arrangements (Poffenberger 2006). Much more forested land is informally managed by communities across the region and their role is likely to become increasingly recognised and empowered as decentralisation devolves natural resource management responsibilities to the local scale. The increasing popularity of community forestry represents a substantial shift in perception as 'government foresters, who once bitterly criticised forest-dependent people as the root cause of deforestation, now view them as the best method to restore and protect watersheds and forestlands' (Poffenberger 2006: 64). Community forestry is not unproblematic, however, as Dressler *et al.* (2006) have shown in their study of local national park management on Palawan Island in the Philippines. In their case study poor quality decision-making processes empowered some more than others and exacerbated local inequalities and tensions rather than resolved them. However it is these sorts of initiatives, alongside a broader environmental awareness, that appear to have the best chance of ensuring the livelihoods of forest dwellers, both human and non-human, are sustained into the future.

Aquatic spaces

Aquatic spaces refer to the rivers, seas and oceans of Southeast Asia. At the local scale aquatic spaces provide important sources of food, freshwater, irrigation, transport, waste disposal, washing and a range of cultural activities to neighbouring communities. Some local communities have established floating villages in which people live on boats and are entirely dependent on aquatic spaces for their livelihoods (see Figures 8.1–8.2). At the national scale stakeholders value aquatic spaces for their contributions to national economies, which can be through port development, hydropower, industrial fishing practices and urban freshwater and waste disposal systems. Most large urban settlements in the region are based on the coast or adjacent to large river networks, facilitating their role in domestic and international trade networks and accelerating their social and economic development. The management of these aquatic spaces, like forested spaces, is vital to people's livelihoods and the long-term development of the region.

The most important rivers within the region include the Mekong River, which flows through five different countries and supports 60 million people within its river basin, and the Salween River, which is important to Myanmar and Thailand and supports 10 million people from 13 different ethnic groups (SEARIN 2006; see Figure 8.3). A particular challenge for the equitable development of rivers relates to

Figure 8.1 Chong Khneas floating village, Tonle Sap, Cambodia.

Source: Author

Figure 8.2 Floating health clinic at Chong Khneas, Cambodia.

Source: Author

controversial national scale proposals for large hydropower dams. Large dams are valued by national authorities for their ability to generate electricity that can be consumed by domestic, largely urban communities, or sold to neighbouring countries to raise capital. One of the largest dams in the region is the Hoa Binh Dam in Vietnam, which has played an important role in the economic development of the country by providing 45 per cent of the country's electricity needs (Hirsch 1998). Laos has recently announced plans to build several dams along the Mekong River with the Nam Theun 2 Dam alone expected to raise US$2 billion in sales of electricity to Thailand in its first 25 years of operation. While such proposals make national economic sense they are controversial due to the upstream and downstream human and ecological costs that are incurred. The Hoa Binh Dam flooded a reservoir that forced 50,000–60,000 people, mainly ethnic minority groups, to abandon their homes and seek out the livelihoods elsewhere. The Nam Theun 2 Dam will have a similar impact upon upstream communities where homes, forests and farmland will be inundated. Downstream communities are also affected when dams are built as river flows are altered, affecting freshwater ecologies, fish stocks, fishing livelihoods, irrigation systems,

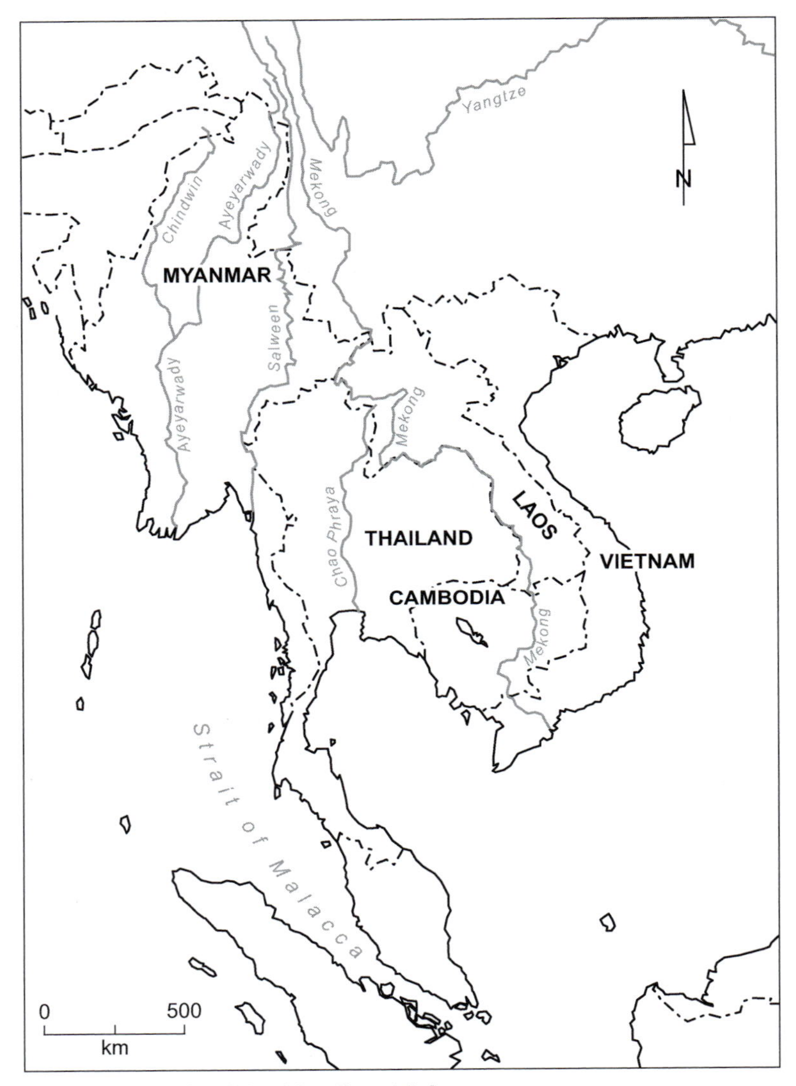

Figure 8.3 Large rivers of mainland Southeast Asia.

river bank gardens, water quality and even lives; children are reported to have drowned from the sudden surges associated with dam openings (see Hirsch 2001). Such changes can be devastating to downstream communities as reflected in the following statement by a member of the Shan minority group in response to plans by the Myanmar government to establish large dams on the Salween: 'If the dam is constructed blocking the river, not only will the Salween River stop flowing but so will Shan history. Our culture will disappear as our houses, temples and farms are flooded' (quoted in Akimoto 2004: 45). Concerns about the local impacts of dams have led to concerted environmental campaigns that draw on regional and international environmental networks to lobby against their construction. This has prevented the building of dams in some places, such as environmentally active Thailand, but has not slowed energy demands leading to the transfer of the damaging impacts of dam-building across borders to Laos and Myanmar. For these poorer countries with suppressed civil society movements dams provide an important source of revenue for national development plans.

The clash of interests between national scale stakeholders and local stakeholders over the management of rivers is not an easy one to resolve. One solution is to not build dams at all because of the negative impacts they have on local communities. This strategy, however, affects more distant communities who would otherwise benefit from electricity supplies or from the investments in social or economic development the state can make with income earned from electricity sales. Indeed many river-based communities are grudgingly accepting of dam-building and are prepared to make personal sacrifices for the national good (perhaps a reflection of Asian values) as long as they are allocated a fair distribution of the benefits (Hirsch 1998). The key to large dam conflicts appears to lie with adequately compensating dispossessed communities for their sacrifices. Too often it seems that the majority of benefits from dam-building are captured by national and international elites involved in dam-building, electricity sales and commercial consumption, as well as the urban middle classes who have high household electricity demands. Local communities often receive poor and inadequate compensation for the difficulties they endure, many of which must uproot their villages, abandon their livelihoods and seek new places to live. Many relocated villagers who do not have official land tenure receive no compensation at all. Considerable research is now being invested in designing 'best practice' dam compensation programmes based on principles of empowerment and participation, however adequately compensating for the loss of homelands and place will never be an easy or straightforward process (see Box 8.2).

Natural spaces do not always fit neatly within the contours of territorial borders. Instead they overlap different territories creating challenges for their management as the policies and approaches of one government may affect the ecologies and ecosystems of neighbouring states. This is particularly the case where aquatic spaces such as rivers

Box 8.2

Dams, rivers and relocating Malaysia's Orang Asli

The construction of the Sungai Selangor Water Supply Scheme Phase III (SSP3) Dam provides an example of the difficulties in compensating people who are disadvantaged by development. The SSP3 Dam was built in 2001 to overcome water shortages in the Malaysian capital Kuala Lumpur. The dam, while small by global standards, displaced eighty-four families in the indigenous *Orang Asli* communities of Peretak and Gerachi. Although similar compensation packages were implemented to restore the living standards of these communities, the outcome of the Peretak resettlement has generally been viewed much more favourably by residents than those in Gerachi.

Prior to their relocation, families in the Peretak and Gerachi villages lived in houses made from bamboo, palm leaves, and in some cases wooden slats and corrugated iron. As part of the compensation package three-bedroom concrete houses with electricity, water and sanitation facilities were provided to each household in both villages. Interviews conducted with community members revealed that the new houses and facilities have proved popular, particularly access to electricity, which allows for the use of television sets, stereos, fridges and electric rice cookers. Other major compensation benefits included cash payments, the plantation of cash crops (oil palm), and the introduction of community development schemes that provided village representatives with job training and employment in the construction of the SSP3 Dam.

While these components of the compensation scheme were generally seen positively it was the location of the resettled communities, and cultural opportunities they afforded, which determined overall community satisfaction. Peretak village was relocated 50 metres from its original position at the edge of the Luit River. As it turned out, the predicted flooding of the original village location did not occur as reservoir levels were lower than expected. The residents of Peretak village enjoy the same place-based environmental values that they did previously and as a result most felt their household's pre-displacement quality of life has been retained or even improved. Most people were happy because their easy access to the river and jungle environment had not be disturbed. As one villager remarked:

> Different houses, money, electricity, they do not matter to us. As long as we have the jungle we can feed our families and make new homes if we want . . . the river provides us with food, water to drink, and a place to wash and swim. The river is especially important for the children, it is their playground. As long as we have these things then we are happy.
>
> (Swainson and McGregor forthcoming)

In Gerachi village, however, the village has moved from the banks of the Selangor River, which is now completely inundated, to a hill approximately 300 metres above its previous setting. Community members are adamant that the involuntary resettlement process has severely diminished their households' quality of life. Respondents spoke of their frustration at living so far from their traditional river environment – something they valued as crucial for their lifestyles. As one Gerachi resident described it: 'We were the guardians of the river, it has always been this way. The children used to play in the river all day, every day. It is now lost to us' (Swainson and McGregor forthcoming). The lessons gleaned from the Peretak and Gerachi case studies suggest that compensation can help minimise the hardships encountered by those adversely affected by development, but it cannot be upheld as the ultimate solution. At the heart of this issue is the fact that the provision of material improvements cannot always replace the intrinsic place-based spiritual and emotional values invested in homeland.

Luke Swainson, Department of Geography, University of Otago

Source: Swainson and McGregor (forthcoming)

and seas flow across and between different territorial jurisdictions. This can contribute to interstate tension and conflict if the actions of one state degrade the water quality of another. To improve the management of these trans-boundary systems the Mekong Rivers Commission (MRC) has been set up. It brings together the governments of Laos, Thailand, Cambodia and Vietnam 'to cooperate in all fields of sustainable development, utilisation, management and conservation of the water and related resources of the Mekong River Basin' (MRC 1995). By providing a forum for inter-governmental discussion, research and negotiation the potential negative impacts of one country's decision to build a dam or release pollutants can hopefully be minimised for downstream states. The MRC, for example, is currently exploring integrated basin flow management techniques to model how dam-building effects downstream biophysical, economic and livelihood qualities (see Guttman 2006). Such institutions can help identify the negative impacts and risks of dams to inform better and more equitable forms of development. However the reach of this international body only stretches so far and it has been unable to prevent China from recently announcing its plans to build eight dams on the Lancang River that flows into the Mekong. China has also announced plans to build thirteen dams on the upper reaches of the Salween having serious downstream consequences for the sustainability and health of two of Southeast Asia's most important rivers.

Another important trans-boundary institution responsible for the management of aquatic spaces is the Southeast Asian Fisheries Development Centre (SEAFDC), which seeks to promote sustainable fisheries and ensure food security for Southeast Asian people through the provision of information, support, advice and training (Ekmaharaj 2007). Fish consumption is a 'way of life' for many Southeast Asian communities, comprising 50 per cent of all animal protein consumed by Indonesians and providing 23.4 kg in the average annual diet of all Southeast Asians (Platon *et al.* 2007; Stobutzki *et al.* 2006). The majority of fishing is undertaken by small scale enterprises, many of which may only involve a small boat and net, and is vital to many poor and landless rural households who may not have access to other income earning opportunities. Some of these households have benefited from development policies that have sought to upgrade their equipment with outboard motors but they have also faced increasing competition from large commercial operations employing industrial fishing technologies such as trawlers (Ebbers *et al.* 2007). This industrial fishing sector has allowed Southeast Asian nations to become net exporters of fish, with Thailand earning over US$4 billion annually from its fish exports (see Figure 8.4).

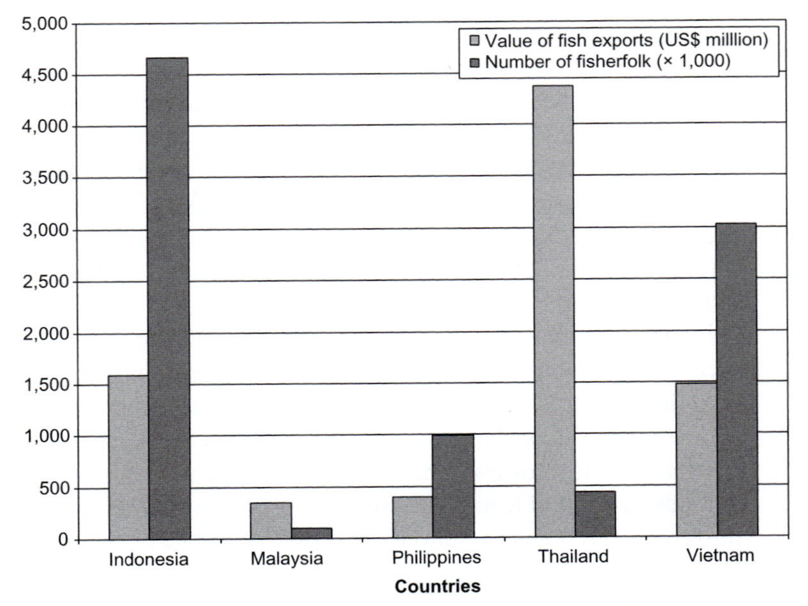

Figure 8.4 Export value and population of fishing industries in select Southeast Asian countries.

Source: Stobutzki *et al.* (2006)

All this activity has put pressure on the sustainability of Southeast Asia's fisheries. The main causes of concern are over-fishing and the degradation of marine ecosystems and habitats through land use changes, pollution, and dynamiting and damage of coral reefs. Some commentators are warning of an imminent collapse in fish stocks and there are already reports of declining fish yields causing hardships for communities that are dependent upon them for food and livelihoods (Stobutzki *et al.* 2006). To combat these emerging challenges SEAFDC is working with Southeast Asian governments to promote aquaculture, size restrictions, quotas, artificial reefs and marine protected areas (Ebbers 2003; Platon *et al.* 2007). While these may seem like top-down approaches there is also a push to decentralise fisheries management so that local fisher-people are involved in decision-making and empowered to seek solutions to what may become a devastating issue (Ebbers *et al.* 2007). As is the case with dams, however, there is a potential clash of interests between states trying to raise revenue through commercial fishing operations and local fisher-people who are struggling to sustain their aquatic livelihoods.

Subsurface spaces

Subsurface spaces refer to natural spaces beneath the earth's surface that are also being altered by activities relating to economic development. This includes processes that are occurring near the surface, such as the changing soil profiles that have accompanied the heavy application of fertilisers, pesticides and herbicides in the wake of the Green Revolution (see Chapter 7). Other processes delve deeper beneath the earth's surface in pursuit of valuable oil, natural gas, coal and mineral deposits. Oil reserves are estimated at 22 billion barrels, natural gas at 227 trillion feet and coal at 46 billion tonnes (ASEAN Secretariat 2004). Oil reserves have historically proved vital for national economic development strategies in Indonesia, Malaysia and Brunei, which collectively produced over one and half million barrels of oil per day in 2006. This translates to multibillion dollar returns with Indonesia and Malaysia both earning close to US$6 billion per annum from oil exports (ASEAN Secretariat 2006). The most important minerals to the region are nickel, copper and tin; bauxite is commonly mined but for small economic returns; and there are a range of gold, lead, zinc, iron ore and gemstone mines, however these minerals are extracted in relatively small volumes. Mineral production only accounted for 0.9 per cent of GDP for the region in 2003, which equates to US$5.9 billion, however some countries, such as Indonesia at 2.3 per cent of GDP, are more reliant than others (ASEAN Secretariat 2005b). International demand for Southeast Asian oil, minerals and energy is expected to grow, particularly given neighbouring China's rapid industrialisation.

Clearly oil, energy and mineral reserves have great value to national scale stakeholders who can earn valuable incomes from extraction, sale and use. They also hold great value to international stakeholders, mainly TNCs, who provide the technical skills, equipment, financial resources and knowledge that is required to set up the mines required to extract subsurface materials. TNCs, as private entities, are becoming increasingly empowered under neo-liberal development paradigms but are primarily concerned with economic profits, which does not often equate with equitable forms of development. Consequently when valuable subsurface deposits are identified there is considerable national and international pressure for the mine to go ahead irrespective of local community interests or consultation. In some situations local communities are happy for mines to be built as it provides them with non-farm employment opportunities and boosts local economic development. In other situations, however, local communities oppose mines because of fears of land loss, congestion, pollution and fundamental changes to their livelihoods. Congestion and lifestyle change accompanied the building of a mine at Teluk Lingga in South Kalimantan that saw 'a sparse settlement of a few scattered houses populated by Banjarese farmers' swell to a population of over 3,000 people in just seven years, causing fundamental shifts in local lifestyles (see Kunanayagam and Young 1998). Similarly a gold mine on Mindanao Island in the Philippines found a local gold mine heightened mercury levels in the air and polluted the local river system causing mercury poisoning and associated illnesses among its workers (Cortes-Maramba *et al.* 2006). If mining companies do not uphold certain environmental standards, which are less stringent and less likely to be enforced in less developed places that are keen on attracting FDI, downstream river pollution can be severe causing 'dead' rivers where no aquatic life can live for miles downstream.

As is the case with large dams, perhaps the biggest concern for local communities is their lack of participation in mining decisions and their associated lack of participation in the distribution of benefits from mining projects. The vast bulk of profits from mining ventures goes towards state and international stakeholders leaving local communities bearing the social and environmental costs for few rewards outside of employment. This has led to outbreaks of frustration among local communities who have protested mine developments, lobbied politicians and damaged mine equipment. The most infamous mine in the region is the Freeport gold mine originally built in the mid-1960s in the province of Irian Jaya in Indonesia. It is the subject of extensive critiques focused upon the environmental damage it has caused to surrounding areas and the lack of benefits or compensation accruing to customary landholders. Protests by local groups have been met with acts of intimidation and violence, some reports indicating that security forces, who have strong links to the Indonesian military, have been involved in acts of murder, torture and rape (see HRW 2007; Soares 2004). Freeport did eventually agree to the establishment of the One Percent Fund, which directed a proportion of company profits to local community development, however this has failed to appease protestors due to its top-down approach, and agitation and violence continue. A lack of participation and involvement in profits accruing from subsurface resources has caused conflicts in other parts of Indonesia, most obviously Aceh and the former province of East Timor. Peluso (2007) has argued that the decentralisation of natural resource management authority to local scale authorities is contributing to further acts of violence as communities compete with one another to assert their claims over contested resources.

A final concern in regards to subsurface materials does not involve conflicts over access, degradation, participation or distribution, but to the unbalanced effect resource dependency can have upon the health and diversity of national economies. Known as the 'resource curse', analysts argue that a comparative advantage in high value natural resources such as oil and minerals can negatively impact the growth of national economies. Some of the reasons for this include:

- an over-reliance on high value natural resources can retard investment in other sectors like manufacturing which has increasing returns over time;
- the shrinkage of non-natural resource sectors makes economies volatile and dependent upon fluctuating world commodity prices;
- Natural resource-based economies encourage corrupt predatory states rather than developmental states as economies do not require government investment and support.

Southeast Asian countries, with the exception of Brunei, have largely avoided the resource curse thesis by successfully diversifying their economies through EOI strategies under strong developmental states.

However the economic growth of China is likely to increase competition for FDI and shrink profitability within manufacturing industries, while also expanding the market for natural resources including oil and minerals. Among the poorer and less established economies of Southeast Asia, including Laos, Myanmar, Vietnam and Cambodia, it is feared that China's growth may push their economies towards the resource curse thesis, further degrading the region's natural spaces (see Coxhead 2006; see Box 8.3).

Box 8.3

Oil: curse or boon?

There is an ongoing debate about the overall impacts the presence of oil and other energy resources, such as natural gas, have upon a country's development. Many believe oil is a key resource that can drive development by generating income from exports while minimising domestic energy costs. Others believe oil to be a curse on development because the resource tends to attract violence, corruption, dependency on an unsustainable capital-intensive resource and environmental damage. Both perspectives apply to different countries in Southeast Asia.

There can be little doubt that those countries in Southeast Asia that have no oil or few oil reserves wish they had more. Singapore, Thailand and the Philippines are all large oil importers that are driving the regional oil trade. Singapore has managed to capitalise on its location and technological efficiencies to benefit from oil economies by becoming a major refining centre and business hub for oil TNCs, as well as an important investor in regional oil companies. Thailand and the Philippines are engaging in offshore oil exploration with some important successes, but their diversified economies are still vulnerable to oil price rises.

Key oil exporters include Brunei, Malaysia, Vietnam and Indonesia. Despite Brunei's wealth it succumbs to the oil curse thesis as its economy has not diversified; oil and gas account for more than 90 per cent of total exports raising questions about the country's long-term economic prospects. Malaysia shows no sign of an oil curse having diversified its economy, benefited from international trade, and experienced few oil-related conflicts, partly because of the high proportion of offshore oil fields. Vietnam has only recently become a significant player in the oil industry and appears to be benefiting economically and developing a diversified economy. Indonesia, which has highest oil production rate in the region, has also diversified its economy but has been embroiled in violent conflicts over access in Aceh and Timor-Leste. In recent years Indonesia has become a net importer of oil as demand outstrips production causing problems for the government who have had to reduce domestic subsidies. This has escalated local oil prices hitting poor households particularly hard and dampened national economic growth.

One country that has suffered the oil curse in the past but is now hoping to use the resource to its maximum advantage is Timor-Leste. The oil curse thesis is evident in Australia's recognition of Indonesia's sovereignty over East Timor after the illegal invasion of 1975. Many believe this recognition, which dampened international efforts to force Indonesia's withdrawal and contributed to years of hardship, was directly linked to oil and gas concessions Indonesia granted Australia in the Timor Sea. Now independent Timor-Leste continues to battle for Timor Sea reserves as Australia has claimed sovereignty over territory not justified under international law and has refused to settle the dispute in the International Court of Law. With few options available and poverty leading to internal unrest, Timor-Leste has agreed to terms with Australia in order to access much needed revenue, but the agreements falls short of provisions outlined within United Nations Laws of the Sea. Timor-Leste is placing revenue from oil and gas extraction into a Petroleum Fund which by 2007 had already attracted US$1 billion. If the fund is used wisely it will diversify the economy and improve the quality of people's lives, if it is not, Timor-Leste may become the latest nation to succumb to the resource curse thesis.

Sources and further reading: La'o Hamutuk (2006, 2007), Clark (2005)

Natural hazards: challenges for development

While natural spaces are usually constructed in terms of how they can contribute to national growth, they also pose considerable risks that can restrain development. Of biggest concern to governments within the region are the threats to life, health, land and property posed by natural hazards. Common hazard risks within Southeast Asia include flooding and landslides, particularly during the monsoon seasons and wild fires and smoke haze during the dry season. Other non-seasonal hazards include earthquakes, volcanoes and, as became tragically evident on Boxing Day 2004, tsunamis. Natural hazards, like natural spaces, result from a combination of non-human and human processes. Nature contributes the unusual conditions, such as tectonic movements, heavy rain or high winds, while human alteration of natural landscapes and their planning and preparedness for extreme events determine the risks natural hazards pose to society. As such 'natural' hazards are only partly natural, and the threats they pose reflect the development decisions of different societies. For example naturally occurring oscillations in river height can result in flooding over river banks, however this is not considered a 'hazard' until it inundates a nearby community. Hence despite a tendency to blame 'nature' for the flooding the hazard itself reflects human decisions to build a community within a flood zone, a lack of flood mitigation protections for that community, and human intensification of the flood event through erosion of river banks and/or upstream land use changes. As such disaster preparedness is becoming an increasingly important component within development planning, something that is necessary from the perspective of equitable development. This is because natural hazards tend to discriminate against the poorest communities as they are the most likely to be living on hazardous land, have the least resilient housing and are the least able to avoid impending hazards. Returning to the flood example those most likely to be inundated are squatter settlements, who have been forced to build on the cheapest and most hazardous flood prone lands, and they are likely to be worst affected because of their poor quality housing. Awareness of the need to mitigate against the threats of natural hazards is growing within development circles, however the costs, expertise and lack of profit-making opportunities have limited concerted action.

In many ways the most unnatural of the 'natural' hazards in Southeast Asia are the fires that regularly sweep through the forests of Borneo, particularly Kalimantan, and Sumatra. They are most common during the dry season when vegetation easily catches fire and rapidly spreads to other areas. While some of these fires have natural origins most are purposefully lit by farmers seeking to clear land for oil palm plantations. Hence the commercialisation of agriculture discussed in Chapter 7 is contributing to 'wild fires' in Indonesia, most of which burn much more than was ever intended. The loss of plant and animal life is immense during these fire events and has ecocentric environmentalists concerned about the survival prospects of endangered species such as the orangutan. Beyond biodiversity concerns are the threats to the life, health and livelihoods of forest dwellers and those in neighbouring settlements who depend upon forested land for their livelihoods. Timber harvesting industries are also affected, with the large Indonesian fires of 1997/1998 estimated to have destroyed timber worth US$2.2 billion from natural and plantation forests (Murdiyarso and Lebel 2007). The health risks posed by fires breach national borders and extend as far as Singapore, Brunei and Malaysia, which are engulfed in seasonal smoke hazes causing respiratory problems as air quality declines well below World Health Organisation safety standards. To mitigate the hazards brought about by fires dozens of international development projects have been implemented over the last 25 years such as the Fire Danger Rating

Systems, which have been set up in Indonesia and Malaysia to provide early warning systems about the potential for serious fire and haze events (de Groot *et al.* 2007). Within the region the ASEAN Agreement on Transboundary Haze Pollution has been signed by several nations and encourages legislation to prevent fires while promoting regional cooperation to minimise haze risks. Unfortunately Indonesia, the source of most fires, has yet to ratify the Agreement and in 2006 the haze was as bad as ever, only dissipating when the monsoon rains commenced in December. The neo-liberal enforced retraction of Indonesian government spending after the Asian economic crisis and the ongoing decentralisation of power may well be diminishing the capacity of the government to respond effectively to fire hazards. Local government may be more reactive to people's needs but they may not have the oversight, coordination, resources and skills required to effectively combat fire threats (see Murdiyarso and Lebel 2007 for a discussion).

Another devastating natural hazard in Southeast Asia derives from severe storm events that cause damage through strong winds, heavy rains and rough seas. Storm events can lead to widespread flooding and landslips that can damage livestock, property, houses and can claim lives. Usually it is poorer communities living on the most hazardous land at the bottom of slopes, beside river banks, or close to the coastline who are the most vulnerable when a storm hits. Even if located in safer locations their properties are less likely to be able to withstand severe storm events making them more vulnerable to hardship. The worst storms are cyclones that approach from the Pacific Ocean with devastating consequences for the Philippines, which experiences over twenty major events per year, but also affecting other countries such as Vietnam and Indonesia. Development strategies in cyclone prone areas seek to minimise the risk people are exposed by empowering them with disaster-mitigation knowledge, investing in comprehensive warning systems as well as encouraging cyclone-proof building standards in local planning regulations. In the Philippines, for example, the 1987 Calamities and Disaster Preparedness Plan provides resources for *barangays* (local regions) to develop hazard contingency plans for cyclones and other events. Disaster networks are kept informed by daily weather bureau updates that detail the strength and damage oncoming winds could cause, providing much needed information for local communities to prepare themselves effectively. Despite these measures, however, cyclones continue to wreak havoc in the Philippines and beyond, partly through a lack of resources to adequately protect vulnerable populations (see Luna 2001). Cyclones can destroy telecommunications and transport infrastructure increasing the risks of hardship and disease once the event has passed.

A further set of hazards emerging from natural spaces derives from the shifting of tectonic plates beneath the earth's surface. Most of Southeast Asia sits on the boundaries of tectonic plates putting countries, particularly Indonesia and the Philippines, at risk of tectonic activity such as volcanoes, earthquakes and tsunamis (see Figure 8.5). Southeast Asia has had the highest number of twentieth-century volcanic incidences of any region in the world with 10,000 Indonesians and 3,000 Filipinos losing their lives, and a further 130,000 and 76,000 respectively being left homeless (Witham 2005). While science and technology have increased the predictability of volcanic eruptions they still pose serious threats. Once again it is the poorest farmers who are the most vulnerable, usually toiling the most dangerous land and being more likely to stay on despite warnings of imminent eruptions due to fears of losing their crops.

Related but less predictable tectonic hazards derive from shifting tectonic plates that cause earthquakes and tsunamis. Earthquakes are relatively frequent events in Indonesia and have devastating consequences. A magnitude 6.3 earthquake occurred in May 2006, for example, killing over 6,000 people in Java and leaving 1.5 million people homeless. Preparedness for earthquakes revolves around construction of earthquake-proof housing but this is expensive and only available for those that can afford it. Tsunamis can be similarly devastating with the Asian or Boxing Day tsunami of 2004 considered one of the worst disasters of all time. Originating off the coast of

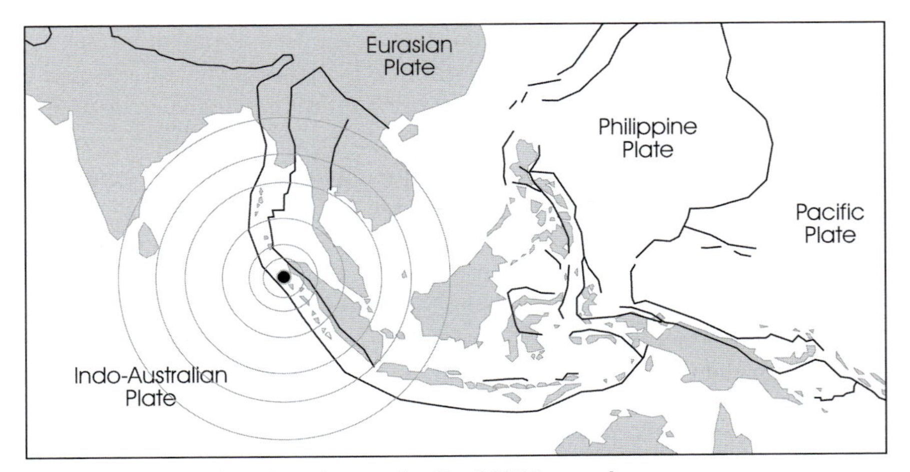

Figure 8.5 Epicentre of earthquake causing the 2004 tsunami.

Sumatra the tsunami devastated the coastal areas of Aceh, including the provincial capital of Banda Aceh, claiming over 160,000 lives. Whole villages were washed away, fishing and farming livelihoods destroyed, and transport and communication infrastructure irreversibly damaged. The tsunami also hit the southwestern coast of Thailand, Myanmar, Bangladesh, Sri Lanka and India severely damaging coastal settlements and shaking national economies. A huge international relief effort swung into action once the scale of the disaster became known, which was clearly beyond the relief capacity of most of the affected states. Rebuilding of homes and other physical infrastructure continues today, however the long-term emotional and psychological damage inflicted upon individuals and communities can never be repaired. It is hoped that early warning systems and education and awareness programmes will lessen the human impacts of a tsunami if one were to occur again (see Box 8.4).

Box 8.4

Vulnerability and disaster: the 2004 Asian tsunami

At approximately 8 am on Boxing Day 2004 an earthquake registering between 9–9.3 on the Richter scale, the second largest ever recorded on a seismograph, erupted off the west coast of Sumatra triggering destructive tsunamis across the Indian ocean. The waves devastated coastal communities claiming the lives of over 230,000 people with thousands more left injured or homeless. The worst affected country was Indonesia where 130,000 people are confirmed dead and 40,000 more still missing. Within Southeast Asia Thailand was also devastated losing 8,000 people in and around some of its most popular tourist destinations. Myanmar has only reported 61 deaths, however experts suspect there could be hundreds, if not thousands, more (see Figures 8.6–8.9).

Aceh, the troubled province that has been embroiled in an independence struggle with the Indonesian government for decades, was the worst affected area in Indonesia. Banda Aceh, the coastal capital, was battered by seven consecutive waves that washed away roads, cars, boats and houses and claimed tens of thousands of lives. In surrounding villages not a single building was left standing as fish ponds, livestock and agricultural fields were destroyed. The Asian Development Bank estimates the financial costs of the disaster to be between US$4.5–5 billion dollars and a massive international recovery operation was launched immediately after the disaster. The early focus was on providing basic medical care, food, water and shelter to ward off disease, before turning to house building, repairing infrastructure and rebuilding livelihoods.

Figure 8.6 Tsunami damage on Khao Phi Phi, Thailand, January 2005.

Source: Dennis Viera © 2005 dennisviera.com

Tsunamis impact more vulnerable sections of the community much harder than others. Women, for example, were much more likely to be killed when the tsunami hit than men; in some villages 80 per cent of the victims were female. Possible explanations for this include that they are less likely to be able to swim or climb palm trees for safety, they were more likely to be looking after and protecting their children, and they were often indoors, having little warning when the tsunami hit. The disaster has also made homeless women in communal barracks vulnerable to sexual abuse and harassment, particularly if they are single or their husband has died.

Poorer households have also been disproportionately affected by the disaster. Their cheaper houses had little strength to resist the waves and were likely to be destroyed; they often lost everything as they had few possessions or financial savings invested elsewhere; their lack of capital prevented them from rebuilding their houses or restarting livelihoods forcing them to remain idle; and their

Figure 8.7 Remains of an Acehnese home after the tsunami, January 2007.

Source: Author

Figure 8.8 Rebuilding homes and livelihoods in Aceh, January 2007.

Source: Author

Figure 8.9 Tsunami warning in Phuket, Thailand.

Source: Author

relative lack of education and low status limited their ability to influence and engage in reconstruction processes. In contrast wealthier households had more resilient homes and access to private resources to repair damage and could restart their livelihoods comparatively quickly. Indeed many have benefited from new employment opportunities finding work in the development and reconstruction industry or in associated hospitality and service industries. In 2007 the contrasts between rich and poor are vividly apparent in the number of four-wheel drives parked outside richer households while poorer communities still reside in communal barracks having waited for years for their lives and homes to be rebuilt.

In recognising the particular vulnerabilities of different sections of society reconstruction efforts can attempt to minimise risks in the future. At the international scale this has contributed to the development of an Indian Ocean Tsunami Warning System that informs governments of events that they must then communicate to at-risk populations. At a more local scale plans can be developed that focus on mitigating the vulnerabilities experienced by sectors of society, such as women and the poor. One of the most effective ways forward, however, would seem to lie in more equitable forms of development so that risks and recovery opportunities are shared and managed more evenly throughout the community.

Sources and further reading: ActionAid (2006), Findlay (2005)

Conclusions

Natural spaces are coming under increasing pressure as Southeast Asia develops. While such spaces have always possessed existence values for the non-human and human communities that rely upon them, development is encouraging people to explore the commercial values of such spaces. This leads to conflicts between local scale stakeholders seeking to protect their livelihoods and more distant stakeholders focused upon national economic development. Such conflicts are further complicated by international stakeholders such as environmental NGOs that may promote ecocentric values or TNCs who are much more profit oriented. From an equitable development perspective it is vital that local stakeholders have a voice within any negotiations that take place. This does not mean that natural spaces cannot be utilised for their resources, as resource use can lead to substantial social and economic gains, but it does mean that local people, who often bear the greatest social and environmental costs and inconveniences, should share in the profits generated. In the past this has not always been the case leading to poverty, discontent, protest and conflict among dispossessed communities. The devolution of natural resource management to local authorities raises hopes that decision-makers will engage with local stakeholders through more participatory approaches and there is evidence of this occurring within community forestry and fisheries programmes. However devolution of authority comes with its own risks such as local scale corruption and biases as well as a lack of trained personnel, which can contribute to poor decision-making and exploitation. Hence there appears to be a role for national scale authorities in the development of natural spaces to provide oversight, coordination, skills and knowledge, to improve the quality of the local scale decision-making that is taking place. Currently however national environmental protection agencies are generally underfunded and poorly staffed and struggle to monitor and uphold regional sustainable development goals and guidelines. The result is an ongoing trend of environmental degradation, particularly within the poorer nations of Southeast Asia where environmental degradation is increasingly being concentrated.

Development is seeking to minimise the risks natural hazards pose to Southeast Asian communities but is having uneven effects. Development is improving people's knowledge of natural hazards, enhancing and expanding hazard warning systems, providing access to hazard mitigation technologies and approaches, and improving relief efforts. However these improvements are not equally accessible to all and for some poorer communities, such as urban squatters or dispossessed rural migrants, hazard risks may have increased because they have been forced to move onto hazard prone land. Providing adequate preparedness, support and compensation where required for these marginal groups is an ongoing challenge for natural resource managers and hazard authorities in Southeast Asian countries. While local, national and international NGOs are increasingly recognising and lobbying on behalf of such groups there is much to be done before natural spaces will be effectively harnessed in the name of the poor.

Summary

- **Natural spaces are valued for a range of reasons that includes their ecocentric conservation values, local livelihood values, and their economic transformative values for national development.**
- **These contrasting values lead to conflict and tension within the management of natural spaces that are experiencing widespread ecological degradation.**
- **Natural resource policies are often used by states to deprive ethnic minority groups of their livelihoods and independence in order to force them into more conventional lifestyles.**
- **Poorer communities are more vulnerable to natural hazards and need to be the focus of development planning to minimise risk.**

Discussion questions

1. Discuss the reasons why Southeast Asia has one of the highest rates of deforestation in the world.
2. Discuss the pros and cons of decentralising decisions about the management of natural spaces to local authorities.

3 Discuss the naturalness of natural hazards and the steps being taken to mitigate the risks they pose.

Further reading

For a good range of interesting case studies regarding conflicts in natural spaces see:

Hirsch, P. and Warren, C. (eds) (1998) *The Politics of the Environment in Southeast Asia: Resources and Resistance*. London: Routledge.

Zerner, C. (ed.) (2003) *Culture and the Question of Rights: Forests, Coasts, and Seas in Southeast Asia*. Durham, NC: Duke University Press.

King, V. (ed.) (1998) *Environmental Challenges in South-East Asia*. Richmond: Curzon.

For a historical approach to contemporary concerns see:

Boomgaard, P. (2007) *Southeast Asia: An Environmental History*. Santa Barbara: ABC-Clio.

An excellent examination of the early forestry industry in Myanmar can be found in:

Bryant, R. (1997) *The Political Ecology of Forestry in Burma 1824–1994*. Honolulu: University of Hawaii Press.

Useful websites

A good source of contemporary environmental information on the Mekong region is the Australian Mekong Resource Centre: www.mekong.es.usyd.edu.au

Regional institutions include the Southeast Asian Rivers Network www.searin.org, the MRC www.mrcmekong.org, the Southeast Asian Rivers Commission www.seafdec.org, and the Asia Forest Network www.asiaforestnetwork.org

The key institution overseeing tsunami reconstruction in Indonesia is the Agency of the Rehabilitation and Reconstruction for the Region and Community of Aceh and Nias, which has large volume of information on its website: www.e-aceh-nias.org/home

9 Towards equitable development

Introduction

Southeast Asia has achieved much since it emerged from colonialism in the immediate aftermath of the Second World War. Poverty is in decline and people are gaining access to more opportunities, a greater range of social services and higher incomes than ever before. The region is experiencing an extended period of peace at the international scale and some long-running internal conflicts in Aceh and Mindanao have shown positive signs of resolution. However beneath this veneer of success there remain concerns about the uneven impacts of development that appear to be benefiting some countries, communities and households, much more than others. Many people feel disempowered and marginalised by development, particularly if it has deprived them of their access to land, livelihoods, power or independence. For these people, and for the long-term prospects of the region, it is essential that development becomes more diverse and responsive to different interests and concerns and shows greater sensitivities to the particularities of place. This final chapter will provide an overview of regional progress towards these more inclusive and equitable forms of development.

Changing approaches to development

The changing theories underlying the development industry have empowered different actors within Southeast Asian development at different stages. The earliest theories, such as modernisation and neo-Marxist theories, directed resources and support to post-colonial state institutions that were seen as vital for effective development. States have subsequently played a pivotal role in Southeast Asian development and retain important functions today. In the immediate post-colonial period they initiated a range of nation-building programmes that cemented national borders and built national identities that attempted to bind disparate communities together. They played important roles in successfully moulding and directing market-led economies through ISI and EOI economic development strategies and were particularly influential in state-led economies where they attempted to distribute wealth and services evenly. They continue to have strong roles within the region today, however, their influence is being challenged, and is in some cases waning, as the popularity of neo-liberal and grassroots development theories grows.

Neo-liberal development theories, supported by powerful institutions such as the World Bank, the IMF and the World Trade Organisation, all of which have become increasingly influential in Southeast Asia since the Asian economic crisis, argue that government spending and influence should be retracted so that free market economies can operate. States are portrayed as overly bureaucratic, inefficient and corrupt, stifling the natural efficiencies and equities of market systems and contributing to inequalities rather than reducing them. Southeast Asia has become increasingly influenced by neo-liberal arguments as evidenced in the establishment of AFTA, which pursues free market principles and through the bilateral free trade agreements various countries have pursued with others. In some cases, such as that of Indonesia, Thailand and the Philippines, they have been forced to undergo neo-liberal reforms in order to access desperately needed IMF loans in the midst of the Asian economic crisis. Cambodia and Timor-Leste turned to similar neo-liberal financial institutions for advice and funding when rebuilding their economies from war and foreign occupation, as have Laos and Vietnam since their economic transition. The IMF and World Bank have been less influential in Singapore, Malaysia and Brunei, where economies are strong but pressure from international trading partners and institutions like the World Trade Organisation, as well as domestic enthusiasm for neo-liberal principles, has seen economic reforms in these countries as well. Even isolationist Myanmar, since the domestic uprising of 1988, has sought to open itself to free market forces and has successfully attracted FDI flows from China and India. As a consequence market forces are growing in power and influence as state services are retracted or privatised and their ability to intervene in economies is restricted.

Grassroots development theories are also challenging state authority and are being promoted within the region by civil society organisations that range from large international NGOs such as Oxfam to local community-based associations of just a few individuals. Grassroots approaches focus on building local community capacity, particularly in sectors or spaces where state services are lacking. As such they tend to be critical of states for geographic or ethnic biases, corruption, inefficiency and a lack of awareness, engagement and sensitivity to the particularities of place. Civil society organisations are becoming more common and influential within Southeast Asia and despite professing a strong dislike for the market-oriented approaches, owe much of their popularity to the spread of neo-liberal thought. NGOs, for example, often provide support to local communities in sectors where state services have been retracted, or were never adequately filled, by providing services such as access to clean water and sanitation, health or education facilities. Despite coming from quite different philosophical principles, grassroots development institutions often find support and funding from the same donors and multilateral institutions that promote neo-liberal development theories.

The consequence of these shifts in development theory has been a partial decline in the power and authority of the state. The state is still a very important institution but its authority and sovereignty is being squeezed by external and internal market forces as well as increasingly active civil society institutions and movements. That states are still so powerful among these winds of change reflects the strength and unity of the ASEAN club, the prominence of the Asian values argument that cements state authority, and the comparative success of some states, such as Singapore, Brunei and Malaysia, in bringing economic development and prosperity to their countries. This has dampened calls for political and economic reform and, as is the case in socialist states, enabled the state to contain and direct civil society movements in non-confrontational ways. Yet even in these countries, the trend is still towards less rather than more state control as commercial entities and social movements gain power and influence. These shifts in emphasis have ramifications for the pursuit of equitable development that can be observed in terms of equality and inequality; political freedoms and opportunities; participation and empowerment; and environmental sustainability.

Equality and inequality

Economic growth has successfully propelled millions of people out of poverty, however it has been uneven, favouring some much more than others. As a consequence inequality, the gap between rich and poor, is increasing at regional, national and local scales throughout Southeast Asia. At the regional scale Singapore and Brunei, and to a lesser extent Malaysia, are providing the majority of their citizens with access to wealth, goods and services that are far beyond the reach of neighbouring economies. Singaporeans can expect annual per capita incomes of over US$25,000 and while Malaysian incomes are closer to US$5,000, they are still far in excess of low income economies where average earnings amount to less than US$2 a day (see Table 1.1). This uneven distribution of wealth is replicated within national borders reflecting market forces that favour particular sections of society over others, resulting in class and space-based divisions. Urban centres, particularly capital cities and extended mega-urban regions, host wealthy elites and expanding middle classes who have access to high income earning opportunities and a wide range of high quality goods and services. In contrast poverty is entrenched in many rural spaces where access to basic services and anything more than subsistence incomes can be difficult. At a more local scale inequality is just as apparent with wealthy urban elites building gated settlements complete with shopping centres and golf courses while squatters struggle in illegal, un-serviced, often hazardous, housing nearby. Similarly the lifestyles of wealthy rural landowners are luxurious when compared to those of the landless labourers who work their sprawling plantations.

The consequences of regional scale inequality are significant. Income disparities are contributing to a regional division of labour where knowledge, financial and service-based industries are concentrated in wealthier urban economies while low paying, often environmentally damaging, agricultural and natural resource-based industries retain prominent roles in poorer economies. The situation in secondary manufacturing industries, which have underpinned regional economic success, is less clear. While high-tech manufacturing industries that require large amounts of capital investment and machinery are still located in the original ASEAN-5 economies there is increasing competition from low income economies, particularly Vietnam, for low skill manufacturing industries in sectors such as textiles, footwear and electronics. If low income economies can compete for a greater share of FDI, including intra-regional FDI from Singapore through bilateral and regional free trade agreements, there is potential for some regional economic inequalities to be lessened, albeit marginally. To prevent this from happening some medium income countries have suppressed low income wages by controlling labour unions, refusing to introduce minimum rates of pay and importing cheap, temporary, unorganised labour from neighbouring low income economies. Many of these migrants suffer difficult working conditions and are susceptible to exploitation giving weight to arguments that suggest uncontrolled free market competition may be contributing to a 'race-to-the-bottom' rather than a rise to the top. Downward pressures are likely to be further exacerbated by low wages in China and its rise as a substantial global competitor for FDI.

At the national scale the high rates of inequality that have evolved within market-led economies are beginning to be replicated in socialist states undergoing market-led reforms. Market forces are concentrating economic activity in urban areas that are growing rapidly and attracting high incomes and a range of good quality services. In contrast poorer rural communities are losing access to good quality health care and education, which are attracting less state and collective community investment. This, and large discrepancies in incomes, is contributing to a region-wide mass migration of people from rural to urban spaces. This is most pronounced in ASEAN-5 economies that now boast sprawling megacities and EMRs and is growing in transitional economies where restrictions on internal population movements have been relaxed. Urban authorities have not been able to cope with the influx and the result has been a proliferation of un-serviced squatter settlements and unregulated informal sector occupations in urban centres. While these settlements and occupations have particularly strengths and are vital to the overall function of urban and national economies they are also generally more hazardous, lower paid and less predictable than their formal sector equivalents. To stem urbanisation processes states are seeking to enhance income earning opportunities in rural spaces by improving transport and communications infrastructure and providing subsidies for businesses who wish to relocate their industrial activities outside core urban centres. Rural industrialisation has a long history in socialist states and is now growing in market-led economies, however the capacity of the state to influence the location of industrial activity is being curtailed by neo-liberal development philosophies that believe markets should be free to determine the locations of economic growth.

At the local scale communities are being encouraged by grassroots development institutions to take control of their own futures and build their internal capacity and self-reliance. A wide range of projects, such as the formation of farmer field schools and urban residents associations, has been initiated that focus on enhancing community assets and expanding livelihood opportunities in locally appropriate and respectful ways. Households are also independently pursuing means of overcoming inequality by diversifying income streams through migration and, in rural areas, by seeking out non-farm incomes through de-agrarianisation. Civil society processes and innovative household strategies empower and assist communities in distributing wealth more evenly at the local scale but have limited impact upon broader national and regional inequalities that are increasingly being driven by market forces. It appears that states still have a strong redistributive role to play if macro-economic inequalities are to be overcome.

Political freedoms and opportunities

There is a trend towards greater political freedoms and opportunities in many parts of Southeast Asia. This reflects neo-liberal and grassroots theories whose advocates promote democratisation as a means of enhancing economic and social freedoms. The most significant shift has occurred in Indonesia, which has emerged from the repressive military rule of President Suharto's New Order government to become the third largest democracy in the world. It joins the Philippines, Thailand, Timor-Leste and, to a lesser extent, Cambodia as the region's most democratic countries. In these countries people enjoy the freedom to vote, to form political parties and have access to a relatively free media, all of which enhance their opportunities to engage in development decision-making. However democracies have their own problems and they face considerable challenges in Southeast Asia in terms of openly and effectively representing the interests of minority groups and their appeals of autonomy, as well as overcoming widespread forms of patronage politics that undermine the effectiveness of formal democratic systems. Nevertheless more people currently possess the right to chose and influence who will govern them than at any point in the past.

Political freedoms are still constrained in other parts of Southeast Asia. In Singapore, Brunei and Malaysia development institutions and domestic populations have acquiesced to the suppression of some political rights in return for strong economic development. These three countries, with their own unique brands of semi-democracy and authoritarianism, have economically outperformed their democratic and socialist regional competitors. Their style of governance has been legitimised through the popularisation of the Asian values argument that pursues economic, social and cultural rights over civil and political rights. However there are signs that people are unhappy with current arrangements as evident in the recent elections in Singapore and Malaysia where the ruling parties struggled to gain the majority votes despite suppressing the activities of oppositional parties. In Malaysia, for example, there is growing support for Islamic rather than economic values, diminishing state claims for legitimacy derived from its economic management credentials. In each of these countries people may well agree with the Asian values argument that economic, social and cultural rights are more important than civil and political rights but, at the very least, spaces should be made available where the pros and cons of different approaches from democracies to meritocracies and socialist states, can be openly debated.

Civil and political rights are still suppressed in Laos and Vietnam where economic liberalisation has not been accompanied by a similar relaxation of political ideology, leaving these two countries on hybrid journeys that merge capitalist and socialist principles. Socialist ideologies position the state as a national organisation that collectively represents the wishes of the people and manages the country in their interests. This negates the need for oppositional parties or elections as communities are, in a sense, the same as the state and various channels and opportunities exist for communities to influence official policies and practices. Enhancing political freedoms and opportunities in Laos and Vietnam does not necessarily require a shift to democracy but it does require improvements in current political systems. It is important that people can engage, critique and influence political decisions without fear of persecution, intimidation or retribution. This is not the case at the moment, particularly for some minority groups, and threatens the pursuit of equitable forms of development. However Myanmar remains the most persecutory and repressive Southeast Asian state and has shown little sign of changing since its disastrous flirtation with democracy in the early 1990s. Isolated from many international development agencies by boycotts and protected within

ASEAN by non-interference principles it is difficult to see how and when political reform may occur. With over 55 million people in the country such reforms are desperately needed if Southeast Asia is to continue in its trend of expanding the political freedoms and opportunities of its populations.

Participation and empowerment

The general trend in the region is towards greater public participation and involvement in development decisions that affect their lives. Once again this reflects the fragmentary pressures of neo-liberalism and grassroots development approaches that have strengthened civil society movements and encouraged political decentralisation. Civil society movements empower people by allowing communities to pursue their interests, air their grievances and act collectively on their concerns. In theory more active civil societies should contribute to more even distributions of wealth and resources as marginalised people are able to have their voices heard and incorporated into development decisions. This has yet to unfold at the national scale in Southeast Asia where countries with well-established civil societies (and well-established market systems) such as the Philippines often display higher levels of inequality than their socialist and semi-democratic neighbours where civil society movements (and market forces) are suppressed. At a more local scale, however, there can be little doubt that labour unions, women's movements and religious institutions are empowering their constituents by representing their interests in local decision-making. Civil societies strengthen unconventional non-economic values and institutions and provide spaces in which locally relevant alternative pathways can be pursued. For these reasons the struggles and achievements of civil society activists in empowering communities to participate in determining their development directions is a welcome and encouraging trend.

Opportunities for greater public participation have also been enhanced by decentralisation processes where power and decision-making responsibilities have been devolved to local authorities. By shrinking the distance between affected communities and community-elected authorities it is hoped that decision-makers will be more sensitive and responsive to the particularities of place. This empowers local communities and institutions by providing them with more opportunities and autonomy in pursuing their own forms of development. However decentralisation comes with its own problems, such as local scale corruption and entrenched patron–client relationships, which can enhance rather than diminish local inequalities. Decentralisation can also lead to uneven development as the assets, strengths and weaknesses of different communities become apparent and the redistributive potential of central authorities is lessened. The more effective decentralisation programmes retain a role for central authorities in overseeing the standard of local scale governance that results. This was the case in Vietnam, for example, during the successful implementation of its land reform programme where central government monitored and publicised local scale corruption and breaches in process. Hence decentralisation enhances the potential for local participation and empowerment in development decision-making, however there is still an important role for central government in monitoring and enhancing the quality of the governance systems that eventuate.

Environmental sustainability

There has been some progress towards more environmentally sustainable relationships with nature through the formation of various regional and national bodies oriented towards environmental protection. These institutions, combined with a growing environmental consciousness and fledgling environmental movements provide hope that current unsustainable approaches to nature can be overcome. The ASEAN Vision 2020 takes a positive approach by calling for sustainable development and the establishment of protective environmental mechanisms across the region. If this is to occur, however, states will need to demonstrate a different level of environmental commitment to what is currently on offer. Environmental authorities are currently under-funded and undervalued when compared to income generating industries that contribute to national economic development. The rapid destruction of Indonesia's forests is perhaps the most obvious example of this but soil degradation, air and water pollution and fisheries depletion are other important examples. The lack of commitment and concern is reflected in market and state approaches to nature that focus on the economic transformative values of nature over its more sustainable locally relevant existence values. This leads to environmental conflict as it is often the poorest and most marginalised communities that are forced to bear the burden of a degraded or hazardous environment while more distant urban or international elites enjoy the majority of the economic benefits for environmentally damaging practices. As economic development increases the pressure on natural spaces within Southeast Asia, particularly in low income countries that have few other means of raising revenue, is only likely to increase.

Of crucial importance to the long-term environmental sustainability of the region will be the unknown impacts that decentralisation will have upon the management of natural spaces. Where decentralisation is occurring local authorities are often inheriting the responsibility of natural resource management. This will hopefully encourage greater environmental stewardship because of the closer connection and dependency that decision-makers have to local environments and the communities that rely on them. There have been some innovative and successful new approaches in community forestry and community fisheries, as well as openings for unconventional approaches such as the forest temples of Buddhist monks in Thailand. However decentralisation also potentially dilutes national environmental initiatives and has contributed to conflicts as different stakeholders seek to assert their claims to resources. Corruption, ethnic biases, limited local scale knowledge of ecological processes and the temptation for short-term profit-making within market-oriented economic systems are all concerns for decentralised natural resource management. Once again there appears to be a role for national authorities within decentralised systems to help resolve conflicts and oversee the pursuit of sustainable human–nature relationships.

Towards equitable development?

At the beginning of this book development was defined, among other things, as 'positive change'. The subsequent chapters explored economic, political and social development, as well as development transformations in urban, rural and natural spaces. From the examples described the development that is occurring is clearly not all 'positive change'. Instead some people are benefiting at the expense of others, economies are becoming increasingly unequal, rural people are leaving their farms and migrating in huge numbers to the cities, millions live in squatter settlements and work in difficult conditions, and natural spaces are rapidly deteriorating with devastating consequences for the ecological systems and livelihoods of those that rely upon them. At the same time, however, there are still many examples of 'positive change'.

The region is becoming wealthier with fewer incidences of poverty, access and the quality of services is improving, there are greater political freedoms, stronger civil societies and greater potential for local scale engagement in development processes than ever before. These strengths, particularly the growth in political freedoms and participatory opportunities, may well provide the basis of more culturally sensitive and appropriate forms of development that can assist in overcoming negative trends related to increasing economic inequality. Markets, states and civil societies will all shape the future of the region; in the right combination they can contribute to more equitable forms of development that redefine the meaning of Southeast Asia as a 'successful developing region'.

References

Abuza, Z. (2002) Tentacles of terror: Al Qaeda's Southeast Asian network. *Contemporary Southeast Asia* 24(3): 427–465.

Achard, F., Eva, H. and Stibig, H. (2002) Determination of deforestation rates of the world's humid tropical forests. *Science* 297: 999–1002.

Acharya, A. (1999) Imagined proximities: the making and unmaking of Southeast Asia as a region. *Southeast Asian Journal of Social Science* 27(1): 55–67.

ActionAid (2006) *Tsunami Response: A Human Rights Assessment*. Johannesburg: ActionAid International.

Aerni, P. and Rieder, P. (2000) Acceptance of modern biotechnology in developing countries: a case study of the Philippines. *International Journal of Biotechnology* 2: 115–131.

Aiga, H. and Umenai, T. (2002) Impact of improvement of water supply on household economy in a squatter area of Manila. *Social Science and Medicine* 55: 627–641.

Akimoto, Y. (2004) *The Salween Under Threat: Damming the Longest Free River in Southeast Asia*. Bangkok: Salween Watch and SEARIN.

Akita, T. and Alisjahbana, A. (2002) Regional income inequality in Indonesia and the initial impact of the economic crisis. *Bulletin of Indonesian Economic Studies* 38(2): 201–222.

Alier, J. (2000) International biopiracy versus the value of local knowledge. *Capitalism, Nature, Socialism* 11(2): 59–65.

Appold, S. (2005) The weakening position of university graduates in Singapore's labor market: causes and consequences. *Population and Development Review* 31(1): 85–112.

ASEAN Secretariat (2002) *ASEAN Report to the World Summit on Sustainable Development*. Jakarta: ASEAN Secretariat.

ASEAN Secretariat (2003) *ASEAN Annual Report 2002–2003*. Jakarta: ASEAN Secretariat.

ASEAN Secretariat (2004) *ASEAN Plan of Action for Energy Coordination 2004–2009*. Jakarta: ASEAN Secretariat.

ASEAN Secretariat (2005a) *ASEAN Statistical Yearbook 2005*. Jakarta: ASEAN Secretariat.

ASEAN Secretariat (2005b) *ASEAN Minerals Cooperation Action Plan 2005–2010*. Jakarta: ASEAN Secretariat.

ASEAN Secretariat (2006) *ASEAN Statistical Pocketbook 2006*. Jakarta: ASEAN Secretariat.

Askew, M. (2002) *Bangkok: Place, Practice and Representation*. London: Routledge.

Aung-Thwin, M. (2001) Parochial universalism, democracy Jihad and the orientalist image of Burma: the new evangelism. *Pacific Affairs* 74(4): 478–505.

Bacani, C. (1997) Surviving the slump. *Asiaweek* 28 November: 52–55.

Bahramitash, R. (2005) *Liberation from Liberalization: Gender and Globalization in Southeast Asia*. London: Zed Books.

Bales, K. (1999) *Disposable People: New Slavery in the Global Economy*. Berkeley: University of California Press.

Banzon-Bautista, C. (1989) The Saudi connection: agrarian change in a Pampangan village, 1977–1984. In Hart, G., Turton, A. and White, B. (eds) *Agrarian Transformations: Local Processes and the State in Southeast Asia*. Berkeley: University of California Press, pp. 144–158.

Barber, B. (1995) The Southeast Asian economic experience and prospects. In Kim, Y. (ed.) *The Southeast Asian Economic Miracle*. Edison, NJ: Transaction Publishers, pp. 243–256.

Barlow, C. (2000) *Institutions and Economic Change in Southeast Asia: The Context of Development from the 1960s to the 1990s*. Northampton: Edward Elgar Publishing.

Baswedan, A. (2004) Political Islam in Indonesia: present and future trajectory. *Asian Survey* 44(5): 669–690.

Beard, V. (2005) Individual determinants of participation in community development in Indonesia. *Environment and Planning C: Government and Policy* 23: 21–39.

Beaverstock, J. and Doel, M. (2001) Unfolding the spatial architecture of the East Asian financial crisis: the organizational response of global investment banks. *Geoforum* 32: 15–32.

Beaverstock, J., Smith, R. and Taylor, P. (1999) A roster of world cities *Cities* 16(6): 445–458.

Bebbington, A., Dharmawan, L., Fahmi, E. and Guggenheim, S. (2006) Local capacity, village governance, and the political economy of rural development in Indonesia. *World Development* 34(11): 1958–1976.

Beeson, M. (2004) *Contemporary Southeast Asia: Regional Dynamics, National Differences*. Basingstoke: Palgrave Macmillan.

Bending, T. (2001) Telling stories: representing the anti-logging movement of the Penan of Sarawak. *European Journal of Development Research* 13(2): 1–25.

Bending, T. (2006) *Penan histories: contentious narratives in upriver Sarawak*. Leiden: KITLV Press.

Beresford, M. (2001) Vietnam: the transition from central planning. In Robinson, R. and G. Rodan (eds) *The Political Economy of Southeast Asia: Conflicts, Crises and Change*. Oxford: Oxford University Press, pp. 206–232.

Berner, E. (1997) Opportunities and insecurities: globalisation, localities and the struggle for Urban Land in Manila. *European Journal of Development Research* 9(1): 167–182.

Berner, E. (2000) Poverty alleviation and the eviction of the poorest: towards urban land reform in the Philippines. *International Journal of Urban and Regional Research* 24(3): 554–566.

Bertrand, J. (2004) *Nationalism and Ethnic Conflict in Indonesia*. New York: Cambridge University Press.

Beyrer, C. (2001) Shan women and girls and sex industry in Southeast Asia: political causes and human rights implications. *Social Science and Medicine* 53: 543–550.

Bloom, G. (1998) Primary health care meets the market in China and Vietnam. *Health Policy* 44(3): 233–252.

Boomgaard, P. (2007) *Southeast Asia: An Environmental History*. Santa Barbara: ABC-Clio.

Booth, A. (1999) The social impact of the Asian crisis: what do we know two years on? *Asian-Pacific Economic Literature* 13(2): 16–29.

Borras, S. (2001) State–society relations in land reform implementation in the Philippines. *Development and Change* 32: 531–561.

Borras, S. and Franco, J. (2005) Struggles for land and livelihood: redistributive reform in agribusiness plantations in the Philippines. *Critical Asian Studies* 37(3): 331–361.

Bottomley, R. (2000) *Structural Analysis of Deforestation in Cambodia (with a Focus on Ratanakiri Province, Northeast Cambodia)*. Tokyo: Mekong Watch and the Institute for Global Environmental Strategies.

Bourgeois, R. and Gouyon, A. (2001) From El Nino to Krismon: how rice farmers in Java coped with multiple crisis. In Gerard, F. and Ruf, F. (eds) *Agriculture in Crisis: People, Commodities and Natural Resources in Indonesia 1996–2000*. Richmond: Curzon Press, pp. 301–333.

Bradsher, K. (2006) Vietnam's roaring economy is set for world stage. *New York Times*

Brown, I. (1997) *Economic Change in Southeast Asia c. 1830–1980*. Kuala Lumpur: Oxford University Press.

Bryant, R. (1997) *The Political Ecology of Forestry in Burma, 1824–1994*. Honolulu: University of Hawaii Press.

Buergin, R. (2003) Shifting frames for local people and forests in a global heritage: the Thung Yai Naresuan Wildlife Sanctuary in the context of Thailand's globalisation and modernisation. *Geoforum* 34: 375–393.

Bunnell, T. (1999) Views from above and below: the Petronas Twin Towers and/in contesting visions of development in contemporary Malaysia. *Singapore Journal of Tropical Geography* 20(1): 1–23.

Bunnell, T. (2004) *Malaysia, Modernity and the Multimedia Super Corridor: A Critical Geography of Intelligent Landscapes*. London: RoutledgeCurzon.

Bunnell, T., Sidaway, J.D. and Grundy-Warr, C. (2006) Re-mapping the 'Growth Triangle': Singapore's cross-border hinterland. *Asia Pacific Viewpoint* 47(2): 235–240.

Carroll, J. (2004) Cracks in the wall of separation? The church, civil society, and the state in the Philippines. In Guan, L. (ed.) *Civil society in Southeast Asia*. Singapore: ISEAS, pp. 54–77.

Case, W. (2002) *Politics in Southeast Asia: Democracy or Less*. Richmond: Curzon.

Case, W. (2003) Interlocking elites in Southeast Asia *Comparative Sociology* 2(1): 249–269.

Chakravarty, S. and Roslan, A. (2005) Ethnic nationalism and income distribution in Malaysia. *European Journal or Development Research* 17(2): 270–288.

Chambers, R. (1997) *Whose Reality Counts? Putting the First Last*. London: Intermediate Technology Publications.

Chernov, J. (2003) Plural society revisited: Chinese–indigenous relations in Southeast Asia. *Nationalism and Ethnic Politics* 9(2): 103–127.

Chia, L. (ed.) (2003) *Southeast Asia Transformed: A Geography of Change*. Singapore: Institute of Southeast Asian Studies.

Christie, C. (1996) *A Modern History of Southeast Asia: Decolonisation, Nationalism and Separatism*. London: Tauris Academic Studies.

Christie, C. (2001) *Ideology and Revolution in Southeast Asia 1900–1980*. Surrey: Curzon.

Clark, M. (2005) Southeast Asia: mixed fortunes. *Petroleum Economist* August: 31–35.

Clarke, G. (2001) From ethnocide to ethnodevelopment? Ethnic minorities and indigenous peoples in Southeast Asia. *Third World Quarterly* 22(3): 413–436.

Cornia, G., Jolly, R. and Stewart, F. (1987) *Adjustment with a Human Face*. Oxford: Oxford University Press.

Cornwel-Smith, P. (2005) *Very Thai: Everyday Popular Culture*. Bangkok: River Books.

Cortes-Maramba, N., Reyes, J., Francisco-Rivera, A., Akagi, H., Sunio, R. and Panganiban, L. (2006) Health and environmental assessment of mercury exposure in a gold mining community in Western Mindanao, Philippines. *Journal of Environmental Management* 81: 126–134.

Cowen, M. and Shenton, R. (1996) *Doctrines of Development*. London: Routledge.

Coxhead, I. (2006) A new resource curse? Impacts of China's boom on comparative advantage and resource dependence in Southeast Asia. *World Development* 35(7): 1099–1119.

Cribb, R. and Narangoa, L. (2004) Orphans of empire: divided peoples, dilemmas of identity, and old imperial borders in East and Southeast Asia. *Comparative Studies in Society and History* 46(1): 164–187.

Cronon, W. (1996) The trouble with wilderness; or getting back to wrong nature. In Cronon, W. (ed.) *Uncommon Ground: Rethinking the Human Place in Nature*. New York: W.W. Norton and Company, pp. 69–90.

de Groot, W., Field, R., Brady, M. Roswintiarti, O. and Mohamad, M. (2007) Development of the Indonesian and Malaysian fire danger rating systems. *Mitigation and Adaption Strategies for Global Change* 12(1): 165–180.

Dery, S. (2000) Agricultural colonisation in Lam Dong Province, Vietnam. *Asia Pacific Viewpoint* 41(1): 35–49.

Devasahayam, T., Huang, S. and Yeoh, B. (2004) Southeast Asian migrant women: navigating borders, negotiating scales. *Singapore Journal of Tropical Geography* 25(2): 135–140.

Dewitt, D. and Hernandez, C. (eds) (2003) *Development and Security in Southeast Asia*. Aldershot: Ashgate.

Dick, H. and Rimmer, P. (1998) Beyond the third world city: the new urban geography of Southeast Asia. *Urban Studies* 35(12): 2303–2321.

Diprose, G. and McGregor, A. (forthcoming) Dissolving the sugar fields: land reform and resistance identities in the Philippines. *Singapore Journal of Tropical Geography*.

Dosch, J. (2007) *The Changing Dynamics of Southeast Asian Politics*. Boulder, CO: Lynne Reinner.

Dressler, W., Kull, C. and Meredith, T. (2006) The politics of decentralizing national parks management in the Philippines. *Political Geography* 25(7): 789–816.

Ducoutieux, O., Laffort, J. and Sacklokham, S. (2005) Land policy and farming practices in Laos. *Development and Change* 36(4): 499–526.

Dutt, A. (1996) *Southeast Asia: A Ten Nation Region*. Boston: Kluwer Academic Publishers.

Dwyer, D. (ed.) (1990) *Southeast Asian Development: Geographical Perspectives*. New York: Longman.

Eaton, P. (2005) *Land Tenure, Conservation and Development in Southeast Asia*. London: Routledge Curzon.

Ebbers, T. (2003) Reconciling fishing and environmental protection: resources enhancement strategies for the conservation and management of fisheries. *Fish for the People* 1(3): 17–26.

Ebbers, T., Weerawat, P. and Eiasma-ard, A. (2007) Capacity building for innovative coastal fisheries management: addressing the changing role of fisheries extension and development. *Fish for the People* 5(1): 17–20.

Eder, J. (2006) Gender relations and household economic planning in the rural Philippines. *Journal of Southeast Asian Studies* 37(3): 397–413.

Ekmaharaj, S. (2007) Responsible fishing technologies and sustainable coastal fisheries management in Southeast Asia. *Fish for the People* 5(1): 10–16.

Elias, J. (2005) The gendered political economy of control and resistance on the shop floor of the multinational firm: a case study from Malaysia. *New Political Economy* 10(22): 203–222.

Elliot, L. (2004) Environmental challenges, policy failure and regional dynamics in Southeast Asia. In Beeson, M. (ed.) *Contemporary Southeast Asia: Regional Dynamics, National Differences*. London: Palgrave Macmillan, pp. 178–197.

Elmhirst, R. (1999) Space, identity, politics and resource control in Indonesia's transmigration programme. *Political Geography* 18: 813–835.

Elson, R. (1992) International commerce, the state and society: economic and social change. In Tarling, N. (ed.) *The Cambridge History of Southeast Asia Vol 2: The 19th and 20th Centuries*. Cambridge: Cambridge University Press, pp. 131–195.

Emmerson, D. (1984) Southeast Asia: what's in a name? *Journal of Southeast Asian Studies* 15(1): 1–21.

Escobar, A. (1995a) *Encountering Development: The Making and Unmaking of the Third World*. Princeton, NJ: Princeton University Press.

Escobar, A. (1995b) Imagining a post-development era. In Crush, J. (ed.) *Power of Development*. London: Routledge, pp. 211–227.

Evers, H. and Korff, R. (2000) *Southeast Asian Urbanism: The Meaning and Power of Social Space*. New York: St Martin's Press.

Evrard, O. and Goudineau, Y. (2004) Planned resettlement, unexpected migrations and cultural trauma in Laos. *Development and Change* 35(5): 937–962.

FAO (2007) *State of the World's Forests 2007*. Rome: Food and Agriculture Organisation.

FAO/GOI (1990) *Situation and Outlook of the Forestry Sector in Indonesia Jakarta*. Jakarta: Food and Agriculture Organisation and Directorate General of Forest Utilisation, Government of Indonesia.

Fealy, G. (2004) Islam in Southeast Asia: domestic pietism, diplomacy and security. In Beeson, M. (ed.) *Contemporary Southeast Asia: Regional Dynamics, National Differences*. London: Palgrave Macmillan, pp. 15–31.

Fearnside, P. (1997) Transmigration in Indonesia: lessons from its environmental and social impacts. *Environmental Management* 21(4): 553–570.

Felker, G. (2003) Southeast Asian industrialisation and the changing global production system. *Third World Quarterly* 24(2): 255–282.

Felker, G. (2004) Southeast Asian development in regional and historical perspective. In Beeson, M. (ed.) *Contemporary Southeast Asia: Regional Dynamics, National Differences*. London: Palgrave Macmillan, pp. 50–74.

Fernando, P., Polet, G., Foead, N., Ng, L., Pastorini, J. and Melnick, D. (2006) Genetic diversity, phylogeny and conservation of the Javan rhinoceros (Rhinoceros sondaicus). *Conservation Genetics* 7(3): 439–448.

Findlay, A. (2005) Editorial: vulnerable spatialities. *Population, Space and Place* 11: 429–439.

Firman, T. (2004a) New town development in Jakarta Metropolitan Region: a perspective of spatial segregation. *Habitat International* 28: 349–368.

Firman, T. (2004b) Demographic and spatial patterns of Indonesia's recent urbanisation. *Population, Space and Place* 10: 421–434.

Foo, G. and Lim, L. (1989) Poverty, ideology and women export factory workers. In Afshar, H. and Agarwal, B. (eds) *Women, Poverty and Ideology in Asia*. London: Macmillan, pp. 212–233.

Forbes, D. (1996) *Asian Metropolis: Urbanisation and the Southeast Asian City*. Melbourne: Oxford University Press.

Forbes, D. and Cutler, C. (2006) Laos in 2005: 30 years of the people's democractic republic. *Asian Survey* 46(1): 175–179.

Forbes, D. and Thrift, N. (1984) Urbanisation in non-capitalist developing countries: the case of Vietnam. In Bedford, R. (ed.) *Essays on Urbanisation in Southeast Asia and the Pacific*. Christchurch: University of Canterbury Press, pp. 1–28.

Ford, M. and Lyons, L. (2006) The borders within: mobility and enclosure in the Riau Islands. *Asia-Pacific Viewpoint* 47(2): 257–271.

Freedom House (2007) *Freedom of the Press 2007: A Global Survey of Media Independence*. Washington: Freedom House.

Frost, S. (2004) Chinese outward direct investment in Southeast Asia: how big are the flows and what does it mean for the region? *The Pacific Review* 17(3): 323–340.

Gellert, P. (2005) The shifting natures of 'development': growth, crisis, and recovery in Indonesia's forests. *World Development* 33(8): 1345–1364.

Ghee, L. (1989) Reconstituting the peasantry: changes in landholding structure in the Muda irrigation scheme. In Hart, G., Turton, A. and White, B. (eds) *Agrarian Transformations: Local Processes and the State in Southeast Asia*. Berkeley: University of California Press, pp. 193–212.

Ghee, L. and Dorall, R. (1992) Contract farming in Malaysia with special reference to FELDA land schemes. In Glover, D. and Ghee, T. (eds) *Contract Farming in Southeast Asia*. Kuala Lumpur: University of Malaya, pp. 71–119.

Gibson, K., Law, L. and McKay, D. (2001) Beyond heroes and victims: Filipina contract migrants, economic activism and class transformations. *International Feminist Journal* 3(3): 365–386.

Gibson-Graham, J.K. (2005) Surplus possibilities: postdevelopment and community economies. *Singapore Journal of Tropical Geography* 26(2): 119–126.
Glassman, J. (2005) The 'war on terrorism' comes to Southeast Asia. *Journal of Contemporary Asia* 35(1): 3–28.
Glassman, J. and Sneddon, C. (2003) Chiang Mai and Khon Kaen as growth poles: regional industrial development in Thailand and its implications for urban sustainability. *Annals of the American Academy of Political and Social Science* 590: 93–115.
Grandstaff, M. and Srisupan, W. (2004) Agropesticide contract sprayers in Central Thailand: health risk and awareness. *Southeast Asian Studies* 42(2): 111–131.
Grenville, S. (2004) The IMF and the Indonesian crisis. *Bulletin of Indonesian Economic Studies* 40 (1): 77–94.
Griffin, K., Khan, A. and Ickowitz, A. (2002) Poverty and the distribution of land. *Journal of Agrarian Change* 2(3): 279–330.
Grundy-Warr, C. and Yin, E. (2002) Geographies of displacement: the Karenni and the Shan across the Myanmar–Thailand border. *Singapore Journal of Tropical Geography* 23(1): 93–122.
Grundy-Warr, C., Peachey, K. and Perry, M. (1999) Fragmented integration in the Singapore–Indonesia border zone: Southeast Asia's 'Growth Triangle' against the global economy. *International Journal of Urban and Regional Research* 23(2): 304–328.
Guan, L. (ed.) (2004) *Civil society in Southeast Asia*. Singapore: ISEAS Publications.
Guan, Y. (2005) Managing sensitivities: religious pluralism, civil society and inter-faith relations in Malaysia. *The Round Table* 94(382): 629–640.
Guttman, H. (2006) River flows and development in the Mekong River Basin. *Mekong Update and Dialogue* 9(3): 2–4.
Ha, H. and Wong, T. (1999) Economic reforms and the new master plan of Ho Chi Minh City, Vietnam: implementation issues and policy recommendations. *GeoJournal* 49: 301–309.
Hadiz, V. (2004) Decentralisation and democracy in Indonesia: a critique of neo-institutionalist perspectives. *Development and Change* 35(4): 697–718.
Hafner, J. (2000) Perspectives on agriculture and rural development. In Leinbach, T. and Ulack, R. (eds) *Southeast Asia: Diversity and Development*. Upper Saddle River, NJ: Prentice Hall, pp. 133–159.
Hagiwara, Y. (1973) Formation and development of the Association of Southeast Asian Nations. *The Developing Economies* XI(4): 443–465.
Hall, D. (1981) *A History of Southeast Asia*. London: Macmillan.
Hamayotsu, K. (2002) Islam and nation building in Southeast Asia: Malaysia and Indonesia in comparative perspective. *Pacific Affairs* 75(3): 353–375.
Hamilton-Merrit, J. (1994) *Tragic Mountains: The Hmong, the Americans, and the secret wars for Laos, 1942–1992*. Bloomington: Indiana University Press.
Hart, G., Turton, A. and White, B. (eds) (1989) *Agrarian Transformations: Local Processes and the State in Southeast Asia*. Berkeley: University of California Press.
Hefner, R. (2006) Islamic economics and global capitalism. *Society* 44(1): 16–22.
Henderson, V. (2002) Urban primacy, external costs, and quality of life. *Resource and Energy Economics* 24: 95–106.

Hersh, J. (1997) The impact of US strategy: making Southeast Asia safe for capitalism. In Schmidt, J., Hersh, J. and Fold, N. (eds) *Social Change in Southeast Asia*. Harlow: Longman.

Hill, H. (1999) An overview of the issues. In Arndt, W. and Hill, H. (eds) *Southeast Asia's Economic Crisis: Origins, Lessons, and the Way Forward*. Sydney: Allen and Unwin.

Hill, H. (2002) *The Economic Development of Southeast Asia*. Cheltenham: Edward Elgar.

Hill, M. (2000) 'Asian values' as reverse Orientalism: Singapore. *Asia Pacific Viewpoint* 41(2): 177–190.

Hill, R. (2002) *Southeast Asia: People, Lands and Economy*. Sydney: Allen & Unwin.

Hilsdon, A. (2003) What the papers say: representing violence against overseas contract workers. *Violence Against Women* 9: 698–722.

Hindmarsh, R. (2003) Genetic modification and the doubly green revolution. *Society* 40(6): 9–19.

Hirsch, P. (1998) Dams, resources and the politics of environment in mainland Southeast Asia. In Hirsch, P. and Warren, C. (eds) *The Politics of the Environment in Southeast Asia: Resources and Resistance*. London: Routledge, pp. 55–70.

Hirsch, P. (2001) Globalisation, regionalisation and local voices: the Asian Development Bank and rescaled politics of environment in the Mekong Region. *Singapore Journal of Tropical Geography* 22(3): 237–251.

Hirsch, P. and Warren, C. (1998) *The Politics of the Environment in Southeast Asia: Resources and Resistance*. London: Routledge.

Hoey, B. (2003) Nationalism in Indonesia: building imagined and intentional communities through transmigration. *Ethnology* 42(2): 109–126.

HRW (2007) *Out of Sight: Endemic Abuse and Impunity in Papua's Central Highlands*. New York: Human Rights Watch.

Huff, W. (2003) Monetisation and financial development in Southeast Asia before the Second World War. *Economic History Review* LVI(2): 300–345.

Hughes, C. (2003) *The Political Economy of Cambodia's Transition, 1991–2001*. London: RoutledgeCurzon.

Hughes, R. (2003) The abject artefacts of memory: photographs from Cambodia's genocide. *Media, Culture and Society* 25: 23–44.

Hun, K. (2002) Research notes on the making of a 'gated community': a study of an inner city neighbourhood, Jakarta, Indonesia. *Asian Journal of Social Science* 30(1): 97–108.

Jagan, L. (2006) Uneasy lies the crown in Myanmar. *Asia Times Online* 4 April 2006.

Jansen, H., Midmore, D., Binh, P., Valasayya, S. and Tru, L. (1996) Profitability and sustainability of peri-urban vegetable production systems in Vietnam. *Netherlands Journal of Agricultural Science* 44: 125–143.

Jenkins, R. (2004) Vietnam in the global economy: trade, employment and poverty. *Journal of International Development* 16(1): 13–28.

Jeyaratnam, J. (1990) Acute pesticide poisoning: a major global health problem. *World Health Statistics Quarterly* 43(3): 139–144.

Jones, G. (1997) The thoroughgoing urbanisation of East and Southeast Asia. *Asia Pacific Viewpoint* 38(3): 237–249.

Jones, G. (2002) Southeast Asian urbanisation and the growth of mega-urban regions. *Journal of Population Research* 19(2): 119–136.

Jones, H. and Pardthaisong, T. (1999) The impact of overseas labour migration on rural Thailand: regional, community and individual dimensions. *Journal of Rural Studies* 15(1): 35–47.

Kadir, S. (2004) Mapping Muslim politics in Southeast Asia after September 11. *The Pacific Review* 17(2): 199–222.

Kay, C. (ed.) (2006) Raul Prebische. In Simon, D. (ed.) *Fifty Key Thinkers on Development*. New York: Routledge, pp. 199–204.

Kelly, P. (2002) Spaces of labour control: comparative perspectives from Southeast Asia. *Transactions: Institute of British Geographers* 27: 395–411.

Kerlvliet, B. and Porter, D. (1995) *Vietnam's Rural Transformation*. Boulder, CO: Westview Press.

King, P. (2004) *West Papua and Indonesia since Suharto: Independence, Autonomy or Chaos?* Sydney: University of New South Wales Press.

King, V. (ed.) (1998) *Environmental Challenges in South-East Asia*. Richmond: Curzon.

King, V. and Kim, Y. (2005) 'Reflexive' reactions of eastern Indonesian women to the economic crisis: an ethnographic study of Tomohon, Minahasa, north Sulawesi. *Indonesia and the Malay World* 33(97): 307–325.

Kingsbury, D. (2001) *South-East Asia: A Political Profile*. Melbourne: Oxford University Press.

Kingsbury, D. (2005) *South-East Asia: A Political Profile (2nd edition)*. Melbourne: Oxford University Press.

Klein, N. (2000) *No Logo: Taking Aim at the Brand Bullies*. New York: Picador.

Koh, G. and Ling, O. (2004) Relationship between state and civil society in Singapore: clarifying the concepts, assessing the ground. In Guan, L. (ed.) *Civil Society in Southeast Asia*. Singapore: Institute of Southeast Asian Studies, pp. 167–197.

Kolko, G. (1997) *Vietnam: Anatomy of a Peace*. London: Routledge.

Kristiansen, S. and Santoso, P. (2006) Surviving decentralisation? Impacts of regional autonomy on health service provision in Indonesia. *Health Policy* 77: 247–259.

Krongkaew, M. and Kakwani, N. (2003) The growth-equity trade-off in modern economic development: the case of Thailand. *Journal of Asian Economics* 14(5): 735–757.

Kunanayagam, R. and Young, K. (1998) Mining, environmental impact and dependent communities: the view from below in East Kalimantan. In Hirsch, P. and Warren, C. (eds) *The Politics of the Environment in Southeast Asia: Resources and Resistance*. London: Routledge, pp. 139–158.

Kusakabe, K. (2004) Women's work and market hierarchies along the border of Lao PDR. *Gender, Place and Culture* 11(4): 581–594.

Kusakabe, K. (2005) Gender mainstreaming in government offices in Thailand, Cambodia, and Laos: perspectives from below. *Gender and Development* 13(2): 46–56.

Lande, C. (1999) Ethnic conflict, ethnic accommodation, and nation-building in Southeast Asia. *Studies in Comparitive International Development* 33(4): 89–117.

Lang, G. and Chan, C. (2006) China's impact on forests in Southeast Asia. *Journal of Contemporary Asia* 36(2): 167–194.

La'o Hamutuk (2006) The CMATS treaty. *The La'o Hamutuk Bulletin* 7(1): 1–12.

La'o Hamutuk (2007) Timor-Leste's petroleum fund. *The La'o Hamutuk Bulletin* 8(1): 1–9.

Laquian, A. (2005) *Beyond Metropolis: The Planning and Governance of Asia's Mega-urban Regions*. Washington: Woodrow Wilson Centre Press.

Latouche, S. (1993) *In the Wake of the Affluent Society: An Exploration of Post-development*. London: Zed Books.

Law, L. (1998) Local autonomy, national policy and global imperatives: sex work and HIV/AIDS in Cebu City, Philippines. *Asia Pacific Viewpoint* 39(1): 53–71.

Leifer, M. (1995) *Dictionary of the Modern Politics of South-East Asia*. London: Routledge.

Leifer, M. (2001) *Dictionary of the Modern Politics of South-East Asia (3rd Edition)*. London: Routledge.

Leinbach, T. and Smith, A. (1994) Off-farm employment, land and lifecycle: transmigrant households in South Sumatra, Indonesia. *Economic Geography* 70(3): 273–296.

Leinbach, T. and Ulack, R. (eds) (2000) *Southeast Asia: Development and Diversity*. Upper Saddle River, NJ: Prentice-Hall.

Lemonie, J. (2005) What is the actual number of the (H)mong in the world. *Hmong Studies Journal* 6: 1–8.

Liamputtong, P. (2005) Birth and social class: northern Thai women's lived experiences of caesarean and vaginal birth. *Sociology of Health and Illness* 27(2): 243–270.

Lian, K. (2006) *Race, Ethnicity, and the State in Malaysia and Singapore*. Boston: Brill.

Liddle, R. and Mujani, S. (2006) Indonesia in 2005: a new multiparty presidential democracy. *Asian Survey* 46(1): 132–139.

Lim, J. and Douglas, I. (1998) The impact of cash cropping on shifting cultivation in Sabah, Malaysia. *Asia-Pacific Viewpoint* 39(3): 315–326.

Lim, L. (1998) *The Sex Sector: The Economic and Social Basis of Prostitution in Southeast Asia*. Geneva: International Labour Office.

Lim, L. and Stern, A. (2002) State power and private profit: the political economy of corruption in Southeast Asia. *Asian-Pacific Economic Literature* 16(2): 18–52.

Linkie, M., Martyr, D., Holden, J., Yanuar, A., Hartana, A., Sugardjito, J. and Leader-Williams, N. (2003) Habitat destruction and poaching threaten the Sumatran tiger in Kerinci Seblat National Park, Sumatra. *Oryx* 37(4): 464–471.

Luna, E. (2001) Disaster mitigation and preparedness: the case of NGOs in the Philippines. *Disasters* 25(3): 216–226.

Luong, H. (2005) The state, local associations, and alternative civilities in rural northern Vietnam. In Weller, R. (ed.) *Civil Life, Globalisation and Political Change in Asia: Organising between the Family and the State*. London: Routledge, pp. 123–147.

Luong, H. (2006) Vietnam in 2006: economic momentum and stronger state–society dialogue. *Asian Survey* 46(1): 148–154.

Lynch, K. (2005) *Rural–Urban Interaction in the Developing World*. London: Routledge.

McCarthy, J. (2005) Between adat and state: institutional arrangements on Sumatra's forest frontier. *Human Ecology* 33(1): 57–82.

McGee, T. (1967) *The Southeast Asian City: A Social Geography of Primate Cities in Southeast Asia*. London: Bell.

McGee, T. (1991) The emerging desakota regions in Asia: expanding a hypothesis. In Ginsburg, N., Koppell, B. and McGee, T. (eds) *The Extended Metropolis: Settlement Transition in Asia*. Honolulu: University of Hawaii Press, pp. 3–25.

McGee, T. (1995) Metrofitting the emerging mega-urban regions of ASEAN: an overview. In McGee, T. and Robinson, I. (eds) *The Mega-urban Regions of Southeast Asia*. Vancouver: UBC Press, pp. 3–26.

McGee, T. and Firman, T. (2000) Labour market adjustment in the time of Krismon: changes in employment structure in Indonesia, 1997–98. *Singapore Journal of Tropical Geography* 21(3): 316–335.

McGee, T. and Robinson, I. (eds) (1995) *The Mega-urban regions of Southeast Asia*. Vancouver: UBC Press.

McGregor, A. (2005) Geopolitics and human rights: unpacking Australia's Burma. *Singapore Journal of Tropical Geography* 26(2): 191–211.

McGregor, A. (2007) Development, foreign aid and post-development in Timor-Leste. *Third World Quarterly* 28(1): 155–170.

Mackerras, C. (ed.) (1995) *East and Southeast Asia: A Multidisciplinary Survey*. Boulder, CO: Lynne Rienner Publishers.

McKinnon, K. (2005) (Im)mobilisation and hegemony: 'hill tribe' subjects and the 'Thai' state. *Social and Cultural Geography* 6(1): 31–46.

Marten, L. (2005) Commercial sex workers: victims, vectors or fighters of the HIV epidemic in Cambodia? *Asia Pacific Viewpoint* 46(1): 21–34.

Martinez, P. (2002) Islam, constitutional democracy, and the Islamic state in Malaysia. In Guan, L. (ed.) *Civil Society in Southeast Asia*. Singapore: ISEAS, pp. 27–53.

Meerman, M. (2001) *Keten van liefde (Chain of Love)*. Brooklyn, NY: Icarus Films.

Menzel, U. (2006) Walt Rostow. In Simon, D. (ed.) *Fifty Key Thinkers on Development*. New York: Routledge, pp. 211–217.

Miller, J. (2003) Why economists are wrong about sweatshops and the antisweatshop movement. *Challenge* 46(1): 93–122.

Milner, A. (1999) What's happened to Asian values? Working paper, Faculty of Asian Studies, Australian National University.

Mkandawire, T. (2007) 'Good governance': the itinerary of an idea. *Development in Practice* 17 (4–5): 679–681.

Momsen, J. (2004) *Gender and Development*. London: Routledge.

Monyrath, N. (2005) The informal economy in Cambodia: an overview. *Economic Review Institute of Cambodia* 1(2): 2–5.

Mosley, L. and Uno, S. (2007) Racing to the bottom or climbing to the top? Economic globalisation and collective labor rights. *Comparative Political Studies* 40: 923–948.

MRC (1995) *Agreement on the Cooperation for the Sustainable Development of the Mekong River Basin 5 April 2005*. Chiang Rai: Mekong River Commission.

Mulder, N. (1997) The legitimacy of the public sphere and the culture of the new urban middle class in the Philippines. In Schmidt, J., Hersch, J. and Fold, N. (eds) *Social Change in Southeast Asia*. Harlow: Longman, pp. 98–113.

Murakami, A., Zain, A., Takeuchi, K., Tsunekawa, A. and Yokota, S. (2005) Trends in urbanisation and patterns of land use in the Asian mega cities Jakarta, Bangkok, and Metro Manila. *Landscape and Urban Planning* 70: 251–259.

Murdiyarso, D. and Lebel, L. (2007) Local to global perspectives on forest and land fires in Southeast Asia. *Mitigation and Adaption Strategies for Global Change* 12(1): 101–112.

Nadvi, K., Thoburn, J., Thang, B., Ha, N., Hoa, N., Le, D. and De Armas, E. (2004) Vietnam in the global garment and textile value chain: impacts on firms and workers. *Journal of International Development* 16(1): 111–123.

Nakagawa, S. (2004) Changing distribution of gender in the Extended Bangkok Region under globalisation. *GeoJournal* 61: 255–262.

Narine, S. (1999) ASEAN into the twenty-first century: problems and prospects. *The Pacific Review* 12(3): 357–380.

Nesadurai, H. (2003) *Globalisation, Domestic Politics and Regionalism: The ASEAN Free Trade Area*. London: Routledge

Noerlund, I. (1997) The labour market in Vietnam: between state incorporation and autonomy. In Schmidt, J., Hersh, J. and Fold, N. (eds) *Social Change in Southeast Asia*. London: Longman, pp. 155–182.

Normille, D. (2007) Indonesia taps village wisdom to fight bird flu. *Science* 315: 30–33.

Ofreno, R.E. and Samonte, I.A. (2005) *Empowering Filipino Migrant Workers: Policy Issues and Challenges*. Geneva: International Labour Organisation Working Paper.

Ong, A. (1987) *Spirits of Resistance and Capitalist Discipline: Factory Women in Malaysia*. Albany, NY: State University of New York Press.

Osbourne, M. (2000) *Southeast Asia: An Introductory History (8th Edition)*. Sydney: Allen and Unwin.

Osbourne, M. (2005) *Southeast Asia: An Introductory History (9th Edition)*. Sydney: Allen and Unwin.

Owen, N. (ed.) (2005) *The Emergence of Modern Southeast Asia: A New History*. Honolulu: University of Hawaii Press.

Peluso, N. (1993) Coercing conservation? The politics of state resource control. *Global Environmental Change* 3(2): 199–217.

Peluso, N. (2007) Violence, decentralisation, and resource access in Indonesia. *Peace Review: A Journal of Social Justice* 19(1): 23–32.

Piper, N. (2004) Rights of foreign workers and the politics of migration in South-East and East Asia. *International Migration* 42(5): 71–97.

Platon, R., Yap, W. and Sulit, V. (2007) Towards sustainable aquaculture in the ASEAN region. *Fish for the People* 5(1): 21–32.

Poffenberger, M. (2006) People in the forest: community forestry experiences from Southeast Asia. *International Journal of Environment and Sustainable Development* 5(1): 57–69.

Porio, E. (2002) Urban poor communities in state-cicil society dynamics: constraints and possibilities for housing and security of tenure in Metro Manila. *Asian Journal of Social Science* 30(1): 73–96.

Porio, E. and Crisol, C. (2004) Property rights, security of tenure and the urban poor in Metro Manila. *Habitat International* 28: 203–219.

Potter, R., Binns, T., Elliot, J. and Smith, D. (2004) *Geographies of Development*. Harlow: Pearson/Prentice Hall.

Powell, B. and Skarbek, D. (2006) Sweatshops and third world living standards: Are the jobs worth the sweat? *Journal of Labor Research* 27(2): 263–274.

Power, M. (2003) *Rethinking development geographies*. London: Routledge.

Power, M. and Sidaway, J. (2004) The degeneration of tropical geography. *Annals of the Association of American Geographers* 94(3): 585–601.

PPI (1967) South East Asia: pulp mill study should be continued. *Pulp and Paper International* 9(13): 8–9.

Prayukvong, W. (2005) A Buddhist economic approach to the development of community enterprises: a case study from Southern Thailand. *Cambridge Journal of Economics* 29: 1171–1185.

Preston, D. (1989) Too busy to farm: under-utilisation of farm land in central Java. *Journal of Development Studies* 26: 43–57.

Pritchard, B. (2006) More than a 'blip': the changed character of South-East Asia's engagement with the global economy in the post-1997 period. *Asia-Pacific Viewpoint* 47(3): 311–326.

Putzel, J. (1999) Survival or an imperfect democracy in the Philippines. *Democratisation* 6(1): 198–223.

Quimpo, N. (2005) The left, elections and the political party system in the Philippines. *Critical Asian Studies* 37(1): 3–28.

Racelis, M. and Aguirre, A. (2002) Child rights for urban poor children in child friendly Philippine cities: views from the community. *Environment and Urbanisation* 14: 97–113.

Rahim, L. (2003) The road less travelled: Islamic militancy in Southeast Asia. *Critical Asian Studies* 35(2): 209–232.

Rajah, A. (1999) Southeast Asia: comparitist errors and the construction of a region. *Southeast Asian Journal of Social Science* 27(1): 41–53.

Ravallion, M. and van de Walle, D. (2004) Breaking up the collective farms: welfare outcomes of Vietnam's massive land privatisation. *Economics of Transition* 12(2): 29–57.

Reid, A. (1980) The structure of cities in Southeast Asia, fifteenth to seventeenth centuries. *Journal of Southeast Asian Studies* 11(2): 235–250.

Reid, A. (1992) Economic and social change 1400–1800. In Tarling, N. (ed.) *The Cambridge History of Southeast Asia Vol. 1: From Early Times to c. 1800*. Cambridge: Cambridge University Press, pp. 460–507.

Rigg, J. (1994) *Southeast Asia: A Region in Transition: A Thematic Human Geography of the ASEAN Region (2nd Edition)*. London: Routledge.

Rigg, J. (1997) Uneven development and the (re-)engagement of Laos. In Dixon, C. and Drakakis-Smith, D. (eds) *Uneven Development in Southeast Asia*. Aldershot: Ashgate, pp. 148–165.

Rigg, J. (1998) Rural–urban interactions, agriculture and wealth: a Southeast Asian perspective. *Progress in Human Geography* 22(4): 497–522.

Rigg, J. (2002) Of miracles and crises: (re-)interpretations of growth and decline in East and Southeast Asia. *Asia Pacific Viewpoint* 43(2): 137–156.

Rigg, J. (2003) *Southeast Asia: The Human Landscape of Modernisation and Development*. London: Routledge.

Rigg, J. (2005) Poverty and livelihoods after full-time farming: a Southeast Asia view. *Asia Pacific Viewpoint* 46(2): 173–184.

Robison, R. (1997) The emergence of the middle class in Southeast Asia and the Indonesian case. In Scmidt, J., Hersh, J. and Fold, N. (eds) *Social Change in Southeast Asia*. New York: Longman.

Rodan, G. (2006) Singapore in 2005: vibrant and cosmopolitan without political pluralism. *Asian Survey* 46(1): 180–186.

Rodan, G., Hewison, K. and Robison, R. (eds) (2001) *The Political Economy of South-East Asia: Conflict, Crises and Change (2nd Edition)*. Melbourne: Oxford University Press.

Rostow, W. (1960) *The Stages of Economic Growth: A Non-communist Manifesto*. Cambridge: Cambridge University Press.

Roth, R. (2004) On the colonial margins and in the global hotspot: park-people conflicts in highland Thailand. *Asia Pacific Viewpoint* 45(1): 13–32.

Routledge, P. (1999) Survival and resistance. In Cloke, P. Crang, P. and Goodwin, M. (eds) *Introducing Human Geographies*. London: Arnold, pp. 76–83.

Rutten, M. (2003) *Rural Capitalists of Asia: A Comparative Analysis on India, Indonesia, and Malaysia*. London: Routledge.

SarDesai, D. (1997) *Southeast Asia: Past and Present (4th Edition)*. Boulder, CO: Westview Press.

Satyanarayana, A. (2002) Birds of passage: migration of South Indian labourers to Southeast Asia. *Critical Asian Studies* 34(1): 89–115.

Savage, V., Kong, L. and Neville, W. (eds) (1998) *The Naga Awakens: Growth and Change in Southeast Asia*. Singapore: Times Academic Press.

Schaumburg-Muller, H. (2005) Private-sector development in a transition economy: the case of Vietnam. *Development in Practice* 15 (3–4): 349–361.

Schumacher, E. (1993) *Small is Beautiful: A Study of Economics as if People Mattered*. London: Vintage.

SEARIN (2006) Southeast Asian Rivers Network. *www.searin.org/index.htm*. Downloaded 15 December 2006.

Shatkin, G. (2004) Planning to forget: informal settlements as 'forgotten places' in globalising Metro Manila. *Urban Studies* 41(12): 2496–2484.

Shiva, V. (2000) *Stolen Harvest: The Hijacking of the Global Food Supply*. Cambridge: South End Press.

Sien, C. (ed.) (2003) *Southeast Asia Transformed: A Geography of Change*. Singapore: ISEAS.

Silvey, R. (2001) Migration under crisis; household safety nets in Indonesia's economic collapse. *Geoforum* 32: 33–45.

Simon, D. (ed.) (2006) *Fifty Key Thinkers on Development*. London: Routledge.

Sioh, M. (1998) Authorizing the Malaysian rainforest: configuring space, contesting claims and conquering imaginaries. *Ecumene* 5(2): 144–166.

Smith, D. (2004) Global cities in East Asia: empirical and conceptual analysis. *International Social Science Journal* 181: 397–412.

Snitwongse, K. and Thompson, W. (2005) *Ethnic Conflicts in Southeast Asia*. Singapore, ISEAS Publications.

Soares, A. (2004) The impact of corporate strategy on community dynamics: a case study of the Freeport Mining Company in West Papua, Indonesia. *International Journal on Minority and Group Rights* 11(1–2): 115–157.

Sobritchea, G. (2002) Women's movements in the Philippines and the politics of critical collaboration with the state. In Guan, L. (ed.) *Civil Society in Southeast Asia*. Singapore: ISEAS, pp. 101–121.

Sodhy, P. (2004) Modernisation and Cambodia. *Journal of Third World Studies* 21(1): 153–174.

Sonnenfeld, D. (1998) Social movements, environment, and technology in Indonesia's pulp and paper industry. *Asia Pacific Viewpoint* 39(1): 95–110.

Sparke, M., Sidaway, J.D., Bunnell, T. and Grundy-Warr, C. (2004) Triangulating the borderless world: geographies of power in the Indonesia–Malaysia–Singapore Growth Triangle. *Transactions of the Institute of British Geographers* 29(4): 485–498.

Steinberg, D. (ed.) (1987) *In Search of Southeast Asia: A Modern History*. Honolulu: University of Hawaii Press.

Stobutzki, I., Silvestre, G. and Garces, L. (2006) Key issues of coastal fisheries in South and Southeast Asia, outcomes of a regional initiative. *Fisheries Research* 78: 109–118.

Sunderlin, W. (2006) Poverty alleviation through community forestry in Cambodia, Laos, and Vietnam: an assessment of the potential. *Forest Policy and Economics* 8(4): 386–396.

Sunderlin, W., Belcher, B., Santoso, L., Angelsen, A., Burgers, P., Nasi, R. and Wunder, S. (2005) Livelihoods, forests, and conservation in developing countries: an overview. *World Development* 33(9): 1383–1402.

Suwanbubbha, P. (2003) Development and Buddhism revisited: arguing the case for Thai religious nuns (Mae Chees). *Development* 46(4): 68–73.

Swainson, L. and McGregor, A. (forthcoming) Development-induced involuntary resettlement: a case study of two Orang Asli communities in Malaysia. *Asia-Pacific Viewpoint*.

Tan, E. (2001) From sojourners to citizens: managing the ethnic Chinese minority in Indonesia and Malaysia. *Ethnic and Racial Studies* 24(6): 949–978.

Tarling, N. (1992a) The establishment of colonial regimes. In Tarling, N. (ed.) *The Cambridge History of Southeast Asia Vol. 2: The 19th and 20th Centuries*. Cambridge: Cambridge University Press, pp. 5–76.

Tarling, N. (1992b) (ed.) *The Cambridge History of Southeast Asia Vol. 1: From Early Times to c. 1800*. Cambridge: Cambridge University Press.

Tarling, N. (1992c) (ed.) *The Cambridge History of Southeast Asia Vol. 2: The 19th and 20th Centuries*. Cambridge: Cambridge University Press.

Taylor, L. (2005) Dangerous trade-offs: the behavioural ecology of child labour and prostitution in rural Northern Thailand. *Current Anthropology* 46(3): 411–423.

Thoburn, J. (2004) Globalisation and poverty in Vietnam: introduction and overview. *Journal of the Asia Pacific Economy* 9(2): 127–144.

Thompson, M. (1996) Late industrialisers, late democratisers: developmental states in the Asia-Pacific. *Third World Quarterly* 17(4): 635–647.

Thorburn, C. (2004) The plot thickens: land administration and policy in post-New Order Indonesia. *Asia Pacific Viewpoint* 45(1): 33–49.

Todaro, M. and Smith, S. (2006) *Economic Development*. Boston: Pearson.

Tomalin, E. (2006) The Thai bhikkhuni movement and women's empowerment. *Gender and Development* 14(3): 385–397.

Transparency International (2006) Corruption perceptions index 2006. *www.transparency.org/policy_research/surveys_indices/cpi/2006* Downloaded 15 September 2007.

Truman, H. (1949) *Inaugural address of the President of the United States*. *www.bartleby.com/124/pres53.html*. Downloaded 13 December 2006.

Un, K. (2006) State, society and democratic consolidation: the case of Cambodia. *Pacific Affairs* 79(2): 225–245.

UNDP (1990) *Human Development Report 1990: Concept and Measurement of Human Development*. New York: United Nations Development Programme.

UNDP (2005) *Human Development Report 2005. International Cooperation at a Crossroads: Aid, Trade and Security in an Unequal World*. New York: United Nations Development Programme.

UNDP (2006) *Human Development Report 2006. Beyond Scarcity: Power, Poverty and the Global Water Crisis*. New York: United Nations Development Programme.

UNDP (2007) *Thailand Human Development Report 2007: Sufficiency Economy and Human Development*. Bangkok: United Nations Development Programme.

UNESCAP (2006) *Statistical Indicators for Asia and the Pacific 2005 Compendium Volume XXXV*. Thailand: United Nations Economic and Social Commission for Asia and the Pacific.

United Nations (2000) *Protocol to Prevent, Supress and Punish Trafficking in Persons, Especially Women and Children (the Trafficking Protocol) Supplementing the United Nations Convention against Transnational Organized Crime*. Geneva: United Nations.

United Nations Human Settlement Programme (2003) *The Challenge of Slums: Global Report on Human Settlements*. London: Earthscan Publications.

Vandergeest, P. and Peluso, N. (1995) Territorialisation and state power in Thailand. *Theory and Society* 24(3) 385–426.

Van Ness, P. (1999) *Debating Human Rights: Critical Essays from the United States and Asia*. London: Routledge.

Vu, T. (2005) Workers and the socialist state: North Vietnam's state–labour relations, 1945–1970. *Communist and Post-Communist Studies* 38(3): 329–356.

Warren, C. (2005) Mapping common future: customary communities, NGOs and the state in Indonesia's reform era. *Development and Change* 36(1): 49–73.

Weightman, B. (2002) *Dragons and Tigers: A Geography of South, East, and Southeast Asia*. New York: John Wiley.

Weller, R. (ed.) (2005) *Civil Life, Globalisation, and Political Change in Asia: Organising between the Family and the State*. London: Routledge.

Wester, L. and Yongvanit, S. (2005) Farmers, foresters and forest temples: conservation in the Dong Mun uplands, Northeast Thailand. *Geoforum* 36: 635–749.

Willis, K. (2005) *Theories and Practices of Development*. London: Routledge.

Winarto, Y. (2004) The evolutionary changes in rice-crop farming: integrated pest management in Indonesia, Cambodia, and Vietnam. *Southeast Asian Studies,* 42(3): 241–272.

Witham, C. (2005) Volcanic disasters and incidents: a new database. *Journal of Vulcanology and Geothermal Research* 148: 191–233.

Woodard, G. (1998) Best practice in Australia's foreign policy: 'Konfrontasi' (1963–66). *Australian Journal of Political Science* 33(1): 85–99.

World Bank (1993) *The East Asian Miracle: Economic Growth and Public Policy*. Oxford: Oxford University Press.

World Bank (2006) *Sustaining Economic Growth, Rural Livelihoods, and Environmental Benefits: Strategic Options for Forest Assistance in Indonesia*. Jakarta: The World Bank Office.

World Bank (2007) The World Bank website. *www.worldbank.org*. Downloaded 1 September 2007.

World Health Organisation (2007a) WHO Avian flu information website. *www.who.int/csr/disease/avian_influenza/en/*. Downloaded 1 September 2007.

World Health Organisation (2007b) WHO SARS information website. *www.who.int/csr/sars/en/*. Downloaded 1 September 2007.

Yah, L. (2004) *Southeast Asia: The Long Road Ahead (2nd Edition)*. Hackensack, NJ: World Scientific.

Yan, L. (2002) Participation of the women's movement in Malaysia: the 1999 general election. In Guan, L. (ed.) *Civil Society in Southeast Asia*. Singapore: ISEAS, pp. 122–143.

Yasmeen, G. (2001) Stockbrokers turned sandwich vendors: the economic crisis and small-scale food retailing in Southeast Asia. *Geoforum* 32: 91–102.

Yeoh, B. (2006) Bifurcated labour: the unequal incorporation of transmigrants in Singapore. *Tijdschrift voor Economische en Sociale Geografie* 97(1): 26–37.

Yeoh, B. and Huang, S. (1998) Negotiating public space: strategies and styles of migrant female domestic workers in Singapore. *Urban Studies* 35: 583–602.

Yuen, B., Yeh, A., Appold, S., Earl, G., Ting, J. and Kwee, L. (2006) High-rise living in Singapore public housing. *Urban Studies* 43(3): 583–600.

Zerner, C. (ed.) (2003) *Culture and the Question of Rights: Forests, Coasts, and Seas in Southeast Asia*. Durham, NC: Duke University Press.

Index

Note: References to boxes, figures and tables are given as *b*, *f*, *t*.